Biosensors

Sensor Physics and Technology Series

Series editors:
Professor K T V Grattan
Centre for Measurement, Instrumentation and Applied Physics
City University
London UK

Dr A Augousti
School of Applied Physics
Kingston University
Kingston-upon-Thames, UK

The *Sensor Physics and Technology Series* aims to bring together in a single series the most important developments in the rapidly-changing area of sensor technology and applications. It will present a snapshot of the range of effort which is being invested internationally in the development of novel types of sensors. New workers in the area of sensor technology will also be catered for with an introduction to the subject through the provision of tutorial guides. Volumes may be sensor technology or applications oriented, and will present recent results from the cutting edge of research in a compact monograph format.

Topics covered will include:

- optical sensors: free-space sensors
- optical sensors: guided wave sensors
- solid state sensors
- biosensors
- microwave sensors
- ultrasonic sensors
- process tomography
- control of networked sensors (system control and data acquisition)
- medical instrumentation
- infrared sensors
- smart sensors
- chemical and biochemical sensing
- environmental sensing
- industrial applications

Titles available

1. Biosensors
   Tran Minh Canh

Biosensors

TRAN MINH CANH

Head of Research
Ecole Nationale Supérieure des Mines
Saint-Etienne, France

Translated by
Sarah A. Jackson, *Ph.D.*

SPRINGER-SCIENCE+BUSINESS MEDIA, B.V.

English language edition 1993

Originally published by Chapman & Hall in 1993
Softcover reprint of the hardcover 1st edition 1993

Original French language edition - *Les Biocapteurs* - © 1991, Masson, Paris

ISBN 978-1-4757-6996-8 ISBN 978-0-585-37623-3 (eBook)
DOI 10.1007/978-0-585-37623-3

A catalogue record for this book is available from the British Library

Library of Congress Cataloging-in-Publication data available

Contents

Preface xi

Abbreviations xiii

1 Introduction 1
1.1 Biosensors - nature's tools? 1
1.2 Historical development 2
1.3 Applications 5

2 General principles 7
2.1 Definitions 7
2.2 General characteristics 7
2.3 Physical biodetection 9
2.3.1 Electrochemical detection 10
(a) Potentiometric techniques 10
(b) Amperometric techniques 11
2.3.2 Thermometric detection 12
2.3.3 Piezoelectric detection 13
2.3.4 Photometric detection 13
2.4 Chemical biodetection 14
2.4.1 Transformation reactions 14
2.4.2 Coupling reactions 15
2.5 Instrumentation 15
2.6 Principles of biosensors 17

3 Construction of biosensors 20
3.1 Choice of bioreceptor 20
3.1.1 Enzymes 20
3.1.2 Microorganisms 21
3.1.3 Tissue and organelles 21
(a) Animal and plant tissue 21
(b) Organelles 21
3.1.4 Immunoreceptors 22
3.1.5 Chemoreceptors 22
3.2 Choice of transducer 22
3.3 Immobilization of bioreceptors 23
3.3.1 Immobilization of enzymes 24
(a) Physical entrapment 24
(b) Immobilization by cross-linking 24
(c) Electromagnetic immobilization 29
(d) Multienzymatic immobilization 29
(e) Immobilization of cofactors 31
(f) Immobilization of mediators 32
(g) Miniature sensors 34
3.3.2 Immobilization of microorganisms 38
3.3.3 Immobilization of immunoagents 39
(a) Immobilization of antibodies 39
(b) Immobilization of antigens 41
(c) Immobilization of compounds with bioaffinity 41
(d) Enzymatic labelling 42
3.3.4 Immobilization of tissue organelles and chemoreceptors 43
(a) Organelles and animal and plant tissue 43
(b) Chemoreceptors 43

4 Enzyme sensors 45
4.1 Principles of operation 45
4.2 Theoretical aspects 46
4.2.1 Response in the transient state 47
4.2.2 Response in the steady state 49
(a) First order kinetics 51
(b) Zeroth order kinetics 51
4.2.3 Stability of enzyme sensors 52
4.2.4 Study of enzymatic inhibition 53

(a) Different types of inhibition 53
(b) Monitoring enzymatic inhibition 57
(c) Enzyme sensor response to different types of inhibitors 58
4.3 Practical aspects 61
4.3.1 Response time 61
4.3.2 Sensor calibration 62
(a) Determination of substrates 62
(b) Determination of inhibitors 63
4.3.3 Parameters affecting 63
(a) Influence of pH 64
(b) Influence of temperature 65
(c) Interference in sensor response 66
4.4 Potentiometric enzyme electrodes 69
4.4.1 Potentiometric urea electrodes 70
(a) pNH_4 transducer 70
(b) pH transducer 71
(c) pCO_2 transducer 71
(d) pNH_3 transducer 73
4.4.2 Potentiometric glucose electrodes 74
4.4.3 Potentiometric amino acid electrodes 75
(a) Oxidases 75
(b) Hydrolases 77
(c) Lyases 77
4.4.4 Other hydrolase electrodes 78
4.4.5 Inhibitor-sensitive enzyme electrodes 80
(a) Reversible inhibitors 81
(b) Irreversible inhibitors 84
4.4.6 Comparison of potentiometric 92
4.5 Amperometric enzyme electrodes 92
4.5.1 Glucose sensitive electrodes 97
(a) Measurement of oxygen partial pressure 98
(b) Measurement of hydrogen peroxide concentration 100
(c) Use of mediators 101
(d) Direct transfer of electrons 103
4.5.2 Determination of polysaccharides 105
4.5.3 Determination of alcohols 106
4.5.4 Determination of lactate 108

4.5.5 Determination of amino acids 109
4.5.6 Comparison of amperometric enzyme electrodes. Multienzyme electrodes 111
4.6 Semiconductor enzyme sensors 116
4.6.1 MOSFET sensors 116
4.6.2 ISFET sensors 117
4.6.3 ENFET sensors 118
4.6.4 Comparison between ENFET sensors and potentiometric enzyme sensors 120
4.7 Optical enzyme sensors 122
4.7.1 Principles of operation 123
4.7.2 Optical sensors based on absorption 124
4.7.3 Optical sensors based on fluorescence 126
(a) Direct measurement of fluorescence 126
(b) Measurement of fluorescence quenching 129
(c) Competitive bonding 131
4.7.4 Optical sensors based on bio/chemiluminescence 132
(a) Bioluminescence 132
(b) Chemiluminescence 135
4.8 Thermal enzyme sensors 136
4.8.1 Principles of thermal enzyme sensors 137
4.8.2 Construction of thermal enzyme sensors 138
(a) Determination of hydrogen peroxide 139
(b) Determination of glucose 139
(c) Determination of urea 140
4.9 Piezoelectric enzyme sensors 142
4.9.1 Principles 142
4.9.2 Construction of piezoelectric sensors 143

5 Microbial sensors 146
5.1 Preparation 146
5.2 Comparison of microbial sensors 146
5.3 Possibilities and limitations 151

6 Immunological sensors 153
6.1 Antigen-antibody coupling 153
6.1.1 Electrochemical sensors 153
6.1.2 Optical sensors 155
(a) Optical fibers 155

(b) Surface plasmon resonance 156
6.1.3 Semiconductor sensors 156
6.1.4 Piezoelectric sensors 157
6.2 Enzymatic labelling 158
6.2.1 Electrochemical methods 158
6.2.2 Optical methods 161
6.2.3 Thermal methods 162

7 Other biosensors 163
7.1 Animal tissue 163
7.2 Plant tissue 163
7.3 Other receptors 164

8 Use and application of biosensors 166
8.1 Operating methods 166
8.1.1 Direct determination in the sample medium. The batch method 166
8.1.2 Flow injection analysis (FIA) 167
(a) Determination of substrates 168
(b) Determination of inhibitors 171
8.1.3 Automation 174
(a) Determination of substrates 174
(b) Determination of inhibitors 174
8.2 Applications 175
8.2.1 The biomedical sector 176
8.2.2 The food produce industry 177
8.2.3 Environmental protection 178
(a) Measurement of organic pollution 178
(b) Measurement of toxicity 179
8.2.4 Defense 179

9 Economic factors 181

10 Conclusions 185

References 187

Index 211

Preface

Of all the recent discoveries in biotechnology, that of biosensor is one of those which has seen an exponential expansion over the last few years. This evolution corresponds with the increasing need for measuring devices that can follow continuously changing biological processes. Biosensors can meet this need provided that their signals include all the information necessary for an understanding of the process, especially concerning the nature and concentration of the species present in the sample medium.

It is well known that sensors form the basis of all instrumental analysis systems, but they also represent the limiting factors of such systems. In this book, we restrict ourselves to the description and study of sensors, leaving aside the different aspects of signal and data treatment. We believe, however, that it is important to stress the multifaceted character of biosensors, and the applications and economic factors which follow.

Biosensor construction is essentially based on the immobilization of a bioreceptor on the corresponding transducer. The reader will find that there are a large variety of techniques for immobilizing enzymes, cofactors and mediators, and even microorganisms, immunoagents, tissues, and organelles. A large part of this book is devoted to enzyme sensors, which is hardly surprising considering that they have been extensively studied and are now commercially available. Other types of biosensors are discussed, with regard to both the principles of their operation, and their construction.

Biosensors form part of the science of engineering in both concept and application. The contents of this book form part of the Biotechnology option at the *Ecole des Mines* in Saint-Etienne, and at the *Institut National Polytechnique* in Grenoble. Although this book is aimed at undergraduate students, the in-depth theoretical and technical discussion may also be useful to graduate students in the fields of biotechnology, biomedicine, analytical chemistry, or instrumentation. The bibliography will also give a research scientist, or a student preparing a thesis, a sound basis and an overview of the field.

The field of biosensors was born out of the collaboration between scientists from a variety of different disciplines: physicists, chemists, medical and pharmaceutical scientists, biologists, electronic engineers, computer scientists, mathematicians, and engineers. All of these scientists will be able to consult this work without any difficulty and find information to complement their original training. Any criticisms or remarks from these readers will be most welcome.

This book is dedicated to the memory of my parents. I am also grateful to my wife Josiane for typing and preparing the manuscript on the word processor. I particularly appreciated her cheerful acceptance of the extra time I needed for editing this book, especially as it occurred so quickly after the publication of the French monograph *Les Biocapteurs*. I also extend my appreciation to my sons Mân and Duc, and to my daughter-in-law Catherine, for their patience and spirit during the preparation of this book.

Abbreviations

AChE	acetylcholinesterase
ADH	alcohol dehydrogenase
ADP	adenosine diphosphate
AFP	alpha-fetoprotein
AMP	adenosine monophosphate
APTES	aminopropyl triethoxysilane
ATP	adenosine triphosphate
BOD	biological oxygen demand
BPT	benzopyrenetetraol
BSA	bovine serum albumin
Con-A	concanavalin A
D-AAO	D-amino acid oxidase
DB18C6	dibenzo-18-crown-6
DIFP	diisopropylfluorophosphonate
DIMP	diisopropylmethylphosphonate
DNP	dinitrophenol
EDTA	ethylene diamine tetraacetic acid
EIA	enzyme immunoassay
ELISA	enzyme linked immunosorbent assay
ENFET	enzyme field-effect transistor
FAD	flavin adenine dinucleotide (oxidized form)
$FADH_2$	flavin adenine dinucleotide (reduced form)
FAPAPP	ferrocenyl amidopentyl amidopropylpyrrole
FAPP	ferrocenyl amidopropylpyrrole
FET	field-effect transistor
FITC	fluorescein isothiocyanate
FMN	flavin mononucleotide (oxidized form)
$FMNH_2$	flavin mononucleotide (reduced form)
GA	glutaraldehyde
GABA	γ-aminobutyric acid
GASFET	gas-sensitive field-effect transistor
GOD	glucose oxidase

HABA	2[(4-hydroxyphenyl)azo] benzoic acid
HCG	human chorionic gonadotropin
HPLC	high pressure liquid chromatography
HPTS	8-hydroxy-1,3,6-pyrenetrisulfonic acid
HSA	human serum albumin
IgG	immunoglobulin G
IMFET	immuno-field-effect transistor
ISE	ion-selective electrode
ISFET	ion-sensitive field-effect transistor
L-AAO	L-amino acid oxidase
LDH	lactate dehydrogenase
LED	light emitting diode
MOS	metal oxide semiconductor
MOSFET	metal oxide semiconductor field-effect transistor
NAD^+	nicotinamide adenine dinucleotide (oxidized form)
NADH	nicotinamide adenine dinucleotide (reduced form)
$NADP^+$	nicotinamide adenine dinucleotide phosphate (oxidized form)
NADPH	nicotinamide adenine dinucleotide phosphate (reduced form)
NMA^+	N-methyl acridinium
NMP^+	N-methyl phenazine
P	inorganic orthophosphate
PAM	pyridine aldoxime-methyl
PMS	phenazine methosulfate
PP	inorganic pyrophosphate
ppb	parts per billion ($1/10^9$)
ppm	parts per million ($1/10^6$)
PTFE	polytetrafluoroethylene
PVC	polyvinylchloride
Q^+	quinolinium
REFET	reference field-effect transistor
SAW	surface acoustic wave
SCE	standard calomel electrode
SPR	surface plasmon resonance
$TCNQ^-$	tetracyanoquinodimethane
TELISA	thermometric enzyme linked immunosorbent assay
$TMPA^+$	trimethylphenylammonium
TPA^+	tetrapentylammonium
TTF^+	tetrathiafulvalene
VDRL	venereal disease research laboratory
XO	xanthine oxidase

1 Introduction

1.1 Biosensors - nature's tools?

Living beings develop and evolve by exchange and communication with their environment. The information gathered enables them to meet their needs, reproduce or survive. Messages to their various senses are received at receptor sites and used to locate food, detect danger, or find a sexual partner. Some bacteria possess specific chemical receptors which, through a positive or negative chemotactism, enable them to locate nutritious substances and distance themselves from poisons.

Man has long dreamt of employing these wonderful tools, biological sensors, to observe and understand his environment. For the moment, however, he must be content with copying nature, and use the arsenal of cells, tissues, proteins, and enzymes that nature has provided to create measuring devices which convert an observed event into a measurable quantity. As a result of this biomimicry, the sensors he creates are given the name *biosensors*. Biosensors not only help him to examine and understand biological processes, but also have numerous industrial and medical applications. The biological medium is continually evolving and thus requires continuous monitoring. This is impossible via traditional sampling methods, and only biosensors can meet such a requirement. Sensors are well known in medicine for the determination of pH, the partial pressure of oxygen or carbon dioxide. Their eventual employment in intensive care or in surgery could give a constant measure of other important biological parameters such as the concentration of glucose, urea, or other metabolites.

1.2 Historical development

In 1962, Clark and Lyons [1] first mentioned the possibility of using enzyme-containing membranes to transform urea or glucose into a product that was detectable with a pH or oxygen electrode. Later, in 1967, Updike and Hicks [2] prepared an enzyme electrode by polymerizing a gel containing glucose oxidase onto an oxygen electrode. When this electrode is placed in contact with a biological solution containing glucose and oxygen, the two compounds diffuse into the enzymatic membrane. The electrode then oxidizes the glucose into gluconic acid using the oxygen in the solution. The oxygen electrode detects the reduction in oxygen partial pressure, which is proportional to the glucose concentration. This discovery was a decisive step in biological analysis. Electrochemistry only provides sensors that detect anions or cations, but by associating them with a biological system, a multitude of other substances can be detected, which would otherwise only be measurable via long and fastidious procedures. Thus, a new class of sensors was created, biosensors. They were born out of the combination of existing sensors, such as amperometric, potentiometric, thermal, piezoelectric, acoustic and optical sensors, and biological systems, such as enzymes, cells, microorganisms, chemical receptors, and immunological agents.

Naturally, the first enzyme fixation methods [2] involved physical retention, and it was attempted to preserve, as far as possible, the integrity of the biological system. It was later realized that covalent fixation would prevent enzyme loss, and ensure long-lasting immobilization. The immobilization of enzymes is of prime importance. So long as the enzymes remain active, their immobilization enables repetitive and multiple determination. Conventional enzymatic methods discard the enzyme after each separate sampling. Enzymes are expensive because of the various extraction and purification stages they require, and we can immediately understand the economic interest of the new procedure, and its impact on the cost of sample analysis in automated systems.

The immobilization of more than one enzyme on a single support was another important step in the development of enzyme sensors [3]. A single enzyme is not always sufficient to transform the substance under study into a substance that is detectable by a transducer. It is often necessary to resort to several enzymes to perform a series of transformations; the product of the first reaction serving as the substrate for the second, and so on [4]. In some cases, the enzyme cofactor, for example, NAD, is not associated with the enzyme and will not be attached to the support during the immobilization of the enzyme. A second reaction must therefore be designed to attach the cofactor to the support, and provide it with the mobility required to reach the active site of the enzyme [5].

The immobilization of mediators has recently been achieved. These compounds, such as ferrocene or the organic conducting salts of tetracyanoquinodimethane (TCNQ), can be immobilized on conducting polymers like polypyrroles [6]. Mediators electrochemically reoxidize the cofactor thereby preventing the dependence of enzyme sensors on the oxygen concentration. In the glucose-sensitive electrode without a mediator, for example, oxygen would be essential for the regeneration of the cofactor FAD during the oxidation of glucose. The use of mediators thus represents an important contribution in biomedical analysis, notably for in-vivo measurements where the oxygen concentration is variable. When mediators are present, the biosensor can easily function in the absence of oxygen, which is vital for the study of anaerobic fermentation processes. The regeneration of mediators, however, requires amperometric electrodes to provide the necessary current for the reoxidation of the cofactor.

Enzyme sensors can also be prepared using field-effect transistors. Such biosensors are constructed by attaching an enzymatic membrane to the insulating layer of a semiconductor structure [7]. Transistors are normally sensitive towards protons, but in the presence of enzymes they may also detect organic substances, such as urea or penicillin. They also have the advantage of a low output impedance, a high signal to noise ratio, and a low price because they can be mass produced. This means that biosensors with a one-time use can be envisaged.

A large technological expansion has also occurred in the area of fiber optics. Improvements have been made in the optical quality and mechanical resistance of optical fibers, and they now have many applications in transducers. Optical fibers lend themselves to the construction of biosensors; a suitable biological substance is attached to the tip of the fiber and produces an optical signal when it is in contact with the system under study [8]. Optical biosensors exploit a variety of radiation phenomena, for example, absorption, fluorescence, and bio/chemiluminescence, and may require single or bifurcated fibers. Optical enzyme sensors are generally used as extrinsic sensors, the fiber is simply used for the transport of light to the sensitive component and then back to the detector.

The technique of detection via light absorption [9] exploits results from UV-visible spectrophotometry, especially in the use of monochromatic radiation and a reference system. This requires the presence of at least two fibers, to separate the outward and inward light beams, and so this method is rarely used. In contrast, fluorescence [10] can be measured directly using a single fiber and very low concentrations can be detected. Furthermore, coupling the immobilized enzyme on an optical fiber with a fluorescent cofactor, such as NADH, opens up greater possibilities. The system is even simpler in the case of bio/chemiluminescence because the reference and excitation beams are obsolete, and the light can be emitted directly to the sensitive component [11]. The excellent quantum yield of bioluminescence also facilitates the detection of low concentrations.

Finally, enzyme biosensors can employ thermal [12] or piezoelectric transducers [13], which detect variations in temperature or mass during an enzymatic reaction. Such biosensors have an extremely rapid response.

Microorganisms, the basic source of enzymatic systems, have also been immobilized on transducers in the place of purified enzymes [14]. This technique avoids the various extraction and purification steps, and keeps the enzyme in its natural environment, which ensures the regeneration of any cofactors. The presence of a basal culture medium is, however, necessary to preserve the enzymatic activity. The specificity of such sensors is limited because of the large number of enzymes present in one microorganism. In contrast, immunological agents [15] are much better adapted for specific recognition of the analyte. Measurement can be performed directly using a variation in some electrical or optical property, a change in mass arising from an antigen-antibody coupling, or via enzymatic labelling, which is analogous to radio-immunological labelling. A number of biosensors employ sections of animal or plant tissue, or even chemical receptors, as biological components [16]. The goal here is to reproduce living detection systems, such as olfactory mucosa, which are well known for their extraordinary specificity and sensitivity.

1.3 Applications

The world is continually confronted by evaluation, and measurement is always the best way of obtaining privileged access to the information essential to scientific progress. Biosensors play an important role in medicine and the development of industrial activities, particularly in automation, regulation, quality control, and energy conservation. In associating a biological component with known transducers, the applications of biosensors can be considerably extended, for example, to biomedical and pharmaceutical fields, and to the food industry, the environment and defense.

The biomedical sector constitutes a potential market for biosensors, especially with the division health care services between hospitals, health care organizations, and the home. Enzyme sensors were therefore the first biosensors to be studied and commercialized for the requirements of biomedical analysis. The first example was the glucose oxidase electrode, which is used to determine glucose in blood and urine for the diagnosis of diabetes. This was followed by the lactate electrode, to evaluate muscular effort, and the urea and creatinin electrodes, to control renal function. Other biosensors use either enzymes or antibodies to determine neurotransmitters, hormones and other metabolites. Implantable biosensors also have applications in the biomedical field, providing access to precise regions of the human body, without consuming, or removing, biological fluids. Implantable biosensors yield immediate results, which are extremely useful if rapid decisions are to be made, for example, during surgery. In the future, biosensors associated with closed-loop systems will help to control and regulate the level of drugs inside the body.

There is a strong demand for biosensors in the food produce industry to follow the various steps in production, and control the quality of the final products. The glucose electrode is used to monitor fermentation, the lactate electrode to control the quality of wines and yoghurt, and the ethanol electrode to evaluate the degree of alcohol in alcoholic drinks.

Microbial electrodes could contaminate the sample medium by the release of microorganisms. They are however suitable for environmental applications, such as estimating organic pollution. They determine biological oxygen demand (BOD), which does not require selective procedures. Total toxicity is measured by monitoring the inhibition of an enzyme immobilized on an electrode. Biosensors can give a continuous indication of the state of our environment, and are good replacements for biological tests, which use living species like fish or daphnia.

Protection against the chemical agents used in warfare remains extremely limited. Here, biosensors represent a major trump card, owing to their low cost, which allows a large number to be used at sensitive sites. Cholinesterase enzyme biosensors reproduce the inhibition of the enzyme in our bodies, and detect the organophosphorus compounds present in most of the chemical warfare agents currently in use. Other biosensors presently under investigation aim at the detection of biological agents, such as the viruses, bacteria and toxic agents that are included in the composition of some bacteriological weapons.

2

General principles

2.1 Definitions

In terms of instrumentation, a *sensor* is defined as a measuring device that exhibits a characteristic of an electrical nature (charge, voltage or current) when it is subjected to a phenomenon that is not electric. The electrical signal it produces must carry all the necessary information about the process under investigation.

Under this definition, a sensor could be regarded as a *transducer*, since it is a system that transforms one physical quantity into another, which is a function of the first. This idea is restrictive because a transducer is a quantitative device, a sensor actually possesses a much larger capacity. First, it recognizes the phenomenon, in a specific way if possible, and translates it into a quantifiable property, which is then transformed into an electrical signal by a transducer. Thus, the original event is represented as an electrical signal and modern techniques of data collection, control and regulation can be applied.

In a *biosensor*, the phenomenon is recognized by a biological system called a *bioreceptor*, which is in direct contact with the sample and forms the sensitive component of the biosensor. The bioreceptor has a particularly selective site that identifies the analyte.

2.2 General characteristics

In order to measure a physical quantity, a sensor must fulfil a number of criteria which indicate its aptitude for meeting the requirement. Biosensors also have to obey this rule. All the useful information about a physical event can be obtained if the biosensor provides a signal that has a direct relationship with the quantity under investigation. The biosensor must also meet requirements connected with the measurement itself; these are repeatability, reproducibility, selectivity, sensitivity, a linear region of response, and a good response time.

Measurements have good *repeatability* if two sets of results obtained by the same operator, using the same sensor in the same sample, are close to each other. The method is *reproducible* if previous results can be obtained by workers in other laboratories. It is essential that a biosensor has high reproducibility if it is to be used commercially.

Biosensors are also characterized by their *specificity*, or their ability to recognize a single compound among other substances in the same sample. Specificity is often impossible to obtain, and so the term *selectivity* is used. A sensor is more selective when the number of interfering compounds is low. The selectivity of biosensors is determined by both the bioreceptor and the method of transduction.

The *sensitivity* of a sensor is given by the change in its response as a function of the corresponding change in the quantity being monitored. Biosensors are more convenient to use if they exhibit a linear relationship between the variation in the amplitude of the output, Δa, and the input, Δm:

$$\Delta a = s\,\Delta m$$

The sensitivity, s, of a biosensor is defined by the ratio $s = \Delta a/\Delta m$, around the measured value, and determines the suitability of the sensor for use in a particular application. It is possible that m is not the actual quantity to be measured, but simply a function of that quantity. This is the case for potentiometric biosensors in which the amplitude of the signal is proportional to the logarithm of the analyte concentration, according to the Nernst Law:

$$\Delta a = s\,\Delta(\log c)$$

When the variation in the phenomenon ceases to yield an appreciable variation in signal, then the *detection limit* has been reached. The detection limit has the same dimensions as the quantity being measured, for most biosensors these are those of concentration. This must not be confused with sensitivity, whose dimensions depend on the type of transducer used. The detection limit usually corresponds to the limit of the linear range at low concentrations. Corrective devices are now available that can go beyond the linear zone and exploit non-linear responses at low concentrations, thus lowering the detection limits of biosensors.

The *linear region* of a biosensor is obtained from a *calibration curve* of its response to different analyte concentrations. The curve is only meaningful if the calibration is made for both increasing and decreasing series of concentrations, and it is important that the biosensor shows no memory effect or hysteresis. A good calibration curve also indicates the stability of the response of the biosensor, which should neither drift nor oscillate with time.

For the measurement to be carried out properly, it is necessary to know the *response time* of the biosensor. This is the time taken to reach a steady state from the instant of the variation in the concentration under investigation. In principle, the response time is infinitely long because of the exponential nature of the curve. A reasonable value can be found, however, by fixing an interval between the instantaneous value of the response and the theoretical final value, as a function of the required precision. The steady state is considered as obtained when this predefined interval has been covered. The response time of the biosensor gives a measure of how quickly it responds to a variation in concentration. It also defines a minimum delay, after which data can be collected with a given precision. The response time is not always used. It is often quicker to obtain the same information from the derivative of the response curve. In systems that use flow injection analysis, only the peak height is taken into account and the achievement of a steady state is unnecessary, also the rate of sample throughput is increased.

2.3 Physical biodetection

A natural event is characterized by certain properties, which can be exploited for its identification using biodetection methods. The senses of living organisms are the most apt tools for the specific recognition of these properties, whose individual features mean that the techniques used for their detection should be based on both the laws of physics and the mechanisms of chemistry. We could say that, out of our five senses, sight, touch and hearing exploit preferentially the principles of physics, whereas smell and taste make more use of chemical reactions. In reality, the information is transmitted through an assembly of physicochemical systems. The same applies for biosensors where, in principle, the transducer plays a physical role, and the bioreceptor has the job of molecular recognition.

The information decoded by the bioreceptor is converted into an electrical signal by the transducer using measuring techniques like potentiometry, amperometry, thermometry, or photometry, all of which are based on the variation of physical quantities. The method chosen must be simple and of a reasonable size, so that it is cheap and easy to use. This is why potentiometric and amperometric electrodes have been so extensively adopted.

2.3.1 Electrochemical detection

(a) Potentiometric techniques

Potentiometry measures the difference in potential between two electrodes immersed in a solution. One of the electrodes probes the solution, while the other serves as a reference. The reference electrode has a constant and reproducible potential which is independent of its environment. The potential of the probe electrode is the potential at the interface between the solid and liquid phases, where the oxidation and reduction reactions occur. For example, at the interface between a conducting wire and a redox system, there is an exchange of electrons between the wire and the compounds being oxidized and reduced. Equilibrium is achieved when the rates of oxidation and reduction are equal, and the composition of the solution surrounding the electrode is constant. The equilibrium potential is then given by the Nernst Law:

$$E = E_0 + \frac{RT}{nF} \ln \frac{[Ox]}{[Red]}$$

where [Ox] and [Red] represent the concentrations of the oxidized and reduced forms of the system, respectively. This form of the Nernst Law is only valid for very dilute solutions. As the ionic concentrations increase, the situation moves away from ideal thermodynamic conditions, and the concentration, c, must be replaced by the activity, a:

$$a = \gamma c$$

where γ is the activity coefficient.

A variety of electrodes are used as a basis for the preparation of potentiometric biosensors. Glass pH electrodes are used for monitoring the concentration of hydrogen ions (pH) or monovalent cations. These are covered with hydrophobic membranes to obtain electrodes which can detect gases (CO_2 or NH_3).

The pH electrode is constructed by attaching a glass membrane with a special composition to the end of a glass tube with high electrical resistance. The internal part of the electrode contains an Ag/AgCl reference system immersed in a buffer solution whose pH is known and fixed. When the electrode is immersed in a solution of unknown pH, the glass membrane only allows the passage of H_3O^+ ions. The electrode potential comes from both the interface potential and the diffusion potential across the glass membrane which acts as an ion exchanger. By modifying the nature of the glass membrane, the electrode can be made sensitive towards monovalent cations such as NH_4^+, Li^+, Na^+ or K^+. These electrodes can then be used to construct enzyme sensors. One example is the urea-sensitive enzyme sensor, in which the NH_4^+ ions produced by the enzymatic reaction are detected by a urease-covered glass membrane.

If the glass membrane of a pH electrode is covered with a gas-permeable hydrophobic membrane, then the electrode is sensitive to gases. An acidic or basic gas (CO_2 or NH_3) can cross the hydrophobic membrane and its presence in the inner solution causes variations in pH which depend upon the gas concentration. A biological component may be attached to this hydrophobic membrane to create a biosensor.

(b) Amperometric techniques

Amperometry is the determination of the intensity of the current crossing an electrochemical cell under an imposed potential. This intensity is a function of the concentration of the electrochemically active species in the sample. Oxidation or reduction of a species is generally performed by a working electrode, and a second electrode acts as a reference. During electrolysis, the working electrode may act as an anode or a cathode, according to the nature of the analyte. For example, a glucose-sensitive biosensor that uses glucose oxidase could detect either the H_2O_2 produced by the enzymatic reaction, or the amount of oxygen consumed during the oxidation of glucose. In the case of H_2O_2, the platinum working electrode is an anode, because it is polarized to a positive potential (+0.6 V/SCE). In contrast, when oxygen is monitored, the working electrode is polarized to a negative potential (–0.65 V/SCE) and is a cathode.

The oxygen probe has a platinum cathode, and a silver anode covered with silver chloride. These are immersed in a solution of KCl, and separated from the analyte sample by an oxygen-permeable membrane. The oxygen diffusing across the membrane is reduced at the cathode, and current produced is proportional to the concentration of oxygen in the sample. A biosensor is obtained by immobilizing a biological system on the plastic oxygen-permeable membrane.

The use of semiconductor structures, such as field-effect transistors (ISFET), means that the results of electrochemical techniques can be exploited, and the problem of the strong impedances exhibited by potentiometric transducers can be solved. The pH and the concentration of ions or even gaseous compounds can be determined by measurement of the drain current in the presence of a reference electrode. Here, the biological system is immobilized on an SiO_2 oxide layer, which forms the sensitive component of the transistor.

2.3.2 Thermometric detection

A biosensor may also use the thermometric detection of the variation in reaction enthalpies. There are many methods of temperature determination (optical, mechanical or electrical). The electrical methods are the most useful for the construction of thermal biosensors because the electrical signal can be obtained directly from the variation in temperature.

Simple thermocouples are not sensitive enough to detect variations in enzymatic reaction enthalpies. Thermistors and thermopiles can, however, detect such small variations in temperature. Thermistors are mixtures of metallic oxides and polycrystalline semiconductors. The high resistivity values of these materials gives a rapid response time owing to their small size and reduced calorific capacity. The resistance R of a thermistor is a function of temperature, T, according to the following expression:

$$R = R_0 \exp B \left(\frac{1}{T} - \frac{1}{T_0} \right)$$

In regions of frequent use, B can be considered as temperature independent.

Thermopiles are made of alternating thermoelectric junctions and are also used to make biosensors. They generate passive signals and thus are particularly suitable for measurements in flowing solutions.

2.3.3 Piezoelectric detection

Piezoelectric measurements use the appearance of an electrical polarization, or a variation in an existing polarization, in certain anisotropic dielectric materials, for example, quartz. This polarization appears when a force is applied in the appropriate direction. The piezoelectric effect is reversible because the material can deform or vibrate when an electric field is applied in the appropriate direction. Piezoelectric devices are used at their resonance frequencies for the determination of small variations in mass. These variations may result from biological reactions that involve association or coupling, for example, enzyme-inhibitor associations or antigen-antibody coupling.

2.3.4 Photometric detection

Optical fibers can also be used in the construction of biosensors. The light between the sample and the source or detector is transported along the interior of the fibers following the principle of total reflection. The light propagates along the fiber in different modes, each corresponding to a given incidence angle. In contrast with interferometric sensors, which require a single mode, biosensors detect variations in light intensity using multimode fibers. Such fibers are the least expensive and have a higher degree of light transmission.

The total reflection in a fiber is never perfect and some electromagnetic radiation penetrates the sheath covering the fiber. This is called the *evanescent wave*, and its intensity diminishes exponentially with the perpendicular distance from the interface. Evanescent waves can sometimes be used to detect variations in optical properties of chemical and biological films placed around the fiber.

A reactive substance is immobilized on an optical fiber, and then the variation in light intensity due to absorption, fluorescence, or luminescence, can be exploited. In the case of absorption, the detector measures the reduction in the intensity of light from the source. This reduction is caused by an absorbent product arising from the reaction between the immobilized substance and the analyte. A larger light intensity is collected than in luminescence. Absorption measurements are generally made using monochromatic wavelength light, and both the incidence and emitted beams have the same wavelength. At least two fibers are necessary, one for the inward beam and one for the outward beam, because the two beam intensities cannot be separated at the same wavelength. The use of white light, however, can yield a complete absorption spectrum with the help of photodiode matrices or arrays.

The wavelength of fluorescence emission is different from its excitation wavelength, and a single fiber is sufficient for transport of both excitation and emission radiation. A single fiber is also used to measure fluorescence quenching, which occurs when a compound that absorbs in the same spectral region as the emission is present. In the future it may be possible to make time-resolved fluorescence measurements. However, fluorescent lifetimes are between 1 and 10 nanoseconds, and such measurements would require complex and costly apparatus.

The most simple photometric method uses the direct detection of the light emitted from chemical and biological changes (chemiluminescence or bioluminescence). These reactions can be catalysed by enzymes immobilized on the end of optical fibers.

2.4 Chemical biodetection

Some biosensors recognize the analyte using transformation and/or coupling reactions.

2.4.1 Transformation reactions

Like all catalysts, enzymes specifically recognize a particular compound, and convert it into a product that is detectable by the transducer. One example of this is penicillinase, which converts the substance under investigation, penicillin, into penicilloic acid, which is then detected using a pH electrode. The intensity of the signal increases with the concentration of penicillin, the analyte.

The substrate of an enzymatic reaction can also be monitored by detecting the consumption of one of its cosubstrates. Glucose can be determined using a pO_2 electrode that detects the oxygen consumption from the enzymatic oxidation of glucose; the signal intensity decreases with glucose concentration. Furthermore, a large number of enzymatic reactions use cofactors, which can be exploited to measure substrate concentration. Ethanol can be determined by fluorimetric detection of the cofactor of alcohol dehydrogenase, NADH, which is immobilized on an optical biosensor.

Finally, it is possible to determine enzyme inhibitors instead of substrates. Inhibitors affect the rates of enzymatic reactions, and hence the intensity of the signal obtained. One example is the cholinesterase electrode which determines inhibitors, especially pesticides.

2.4.2 Coupling reactions

Coupling reactions between antigens and antibodies are generally extremely specific. They can produce variations in electric charge, mass, and optical properties, all of which are directly detectable by transducers. Variations in charge can be amplified by associating the antibody with an ionophore, whose charge is modified by the coupling reaction and can be detected with a potentiometric electrode. Antibodies and the complement system lyse certain cells. This property can be exploited because it yields ions that are detectable with an electrode during an antigenic reaction. Variations in mass upon coupling are easiest to detect with piezoelectric transducers. The detection limit here is very low and so such detectors are especially suitable for monitoring immunoagents or enzyme inhibitors whose concentrations are often very low.

Finally, both coupling and transformation reactions can be detected using enzymatic labelling of immunoagents with, for example, peroxidases. The label ensures that there is sufficient consumption of a substrate, or the generation of a product, for detection by potentiometric, optical or thermal transducers.

2.5 Instrumentation

Most biosensors make use of existing transducers and the instrumentation already associated with them. The principal modifications are made to the part of the transducer where the biological system is to be situated. Biological systems must be renewed periodically to maintain an optimal activity and the membranes that carry proteins and other reactive substances are much easier to handle if they are attached to removable or disposable components.

The choice of biosensor is often related to the cost of the instrumentation for a given application. Biosensors will be used more extensively in health care in the future, provided that their total cost, including the instrumentation, is not excessive. The same applies to the environment and defense, where more biosensors equates with more effective protection. In this respect, electrochemical biosensors appear to be well placed, requiring apparatus that is both simple and small. Some pH electrodes are the size of a pencil (including their amplification and display system), and it is possible to attach a biological component to them.

Optical biosensors generally require much larger apparatus with systems of lenses, mirrors, monochromators, and photomultipliers. The light sources may be incandescent lamps with filters and monochromators. Tungsten filament lamps have an intense light beam but also a high thermal inertia, which makes rapid modulation of the input current impossible. In contrast, light emitting diodes have much shorter response times, which means that their relatively weak light beam can be modulated right up to high frequencies, but the choice of wavelength remains rather limited. Diode lasers have an even higher modulation rate, higher power, and more monochromatic light, but are only available in the infrared region of the spectrum. In some applications, other types of lasers can be used as light sources, for example, solid or gas lasers, or dye lasers. Argon lasers, for example, give extremely strong light intensity and the size of the optical fibers can be reduced.

The performance of optical fibers is mostly related to their cost. Fibers that are made from plastic materials are cheap but are limited to visible radiation. Glass fibers must be used below 450 nm, and for the ultraviolet above 280 nm, quartz fibers are required, which are more expensive and much more fragile than plastic fibers. The signal is detected using photomultipliers, photodiodes, and detectors using arrays or matrices. The choice between the different methods depends on the level of light intensity emitted and may increase the total cost of the analysis considerably.

Thermal biosensors require only very simple apparatus. Thermistors measure the temperature change of an enzymatic reaction by the simple measurement of variation in resistance. In addition to the enzymatic thermistor, it is advisable to have a reference thermistor without an enzyme, which accounts for thermal perturbations. The difference in resistance is measured using a Wheatstone bridge and temperature differences can be detected to a minimum of 10^{-4} °C. Thermopiles based on multiple thermocouple junctions can be used instead of thermistors, and then the output voltage is measured with a microvoltmeter or a recorder.

Piezoelectric biosensors exploit variations in the resonance frequency of a quartz crystal. In order to vibrate the crystal, an oscillator is connected to electrodes placed on both sides of the crystal. The vibrational frequency of the crystal is monitored using a frequency meter, which has a resolution of about 0.1 Hz and displays the results directly. The data can also be recorded potentiometrically using a digital to analog converter.

All biosensors need a recording system to observe the reproducibility of the signal. This also indicates the nature of the response curve, and detects any anomalies. Furthermore, it provides a permanent record of the behavior of the biosensor during both calibration and the determination itself.

Some biosensors were designed to follow the different stages of a production line one by one, by semi-continuous measurement. In this case, the biosensor is used with a system of flow injection analysis (FIA). Small sample volumes are used, and the problems of contamination in the fermentation industry can be avoided. There are also applications in biomedical analysis laboratories, where faster methods reduce the cost of each analysis. Finally, the automation of FIA is possible using a microcomputer for data collection and management of the various steps of biosensor calibration and measurement.

2.6 Principles of biosensors

A biosensor can be considered as a combination of a *bioreceptor*, the biological component, and a *transducer*, the detection method. The total effect of a biosensor is to transform a biological event into an electrical signal. Figure 2.1 represents the principle of the operation of a biosensor which, starting from the analyte, can provide all the information needed for its evaluation. This information can be processed and stored for later use.

The first link of a biosensor is the *bioreceptor*, which has a particularly selective site that identifies the analyte. The bioreceptor ensures molecular recognition, and may transform the analyte in some way. This localized modification is generally made via an immobilized enzyme, which transforms the analyte into a product that is detectable by the transducer. This is the case for enzyme sensors. Sometimes, however, the enzyme is only stable in its natural environment, which cannot be modified, and the whole cell or microorganism is immobilized on the biosensor.

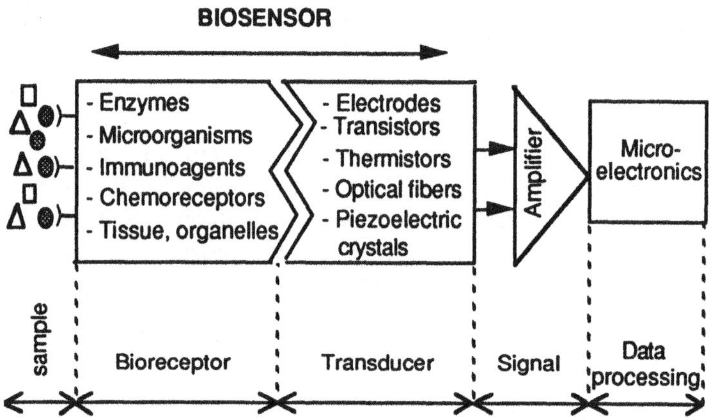

Figure 2.1 *The various stages of determination with a biosensor.*

The other component of a biosensor is the *transducer*, which exploits the biochemical modification of the substrate by the bioreceptor by transforming it into an electrical signal. We could say that the transducer converts one type of energy into another. The choice of transducer depends on the type of biochemical modification, it should make optimal use of the product of the bioreceptor and give a signal that is sensitive, easily monitored, and has minimal background noise. Low background noise reduces the detection limit and improves the biosensor performance.

The combination of any bioreceptor with any transducer leads to a large number of biosensors. In reality, the two components have to be compatible to give rise to an electrical signal. It is impossible, for example, to use a thermometric transducer if the substrate transformation reaction does not give rise to a variation in enthalpy. Electrochemical transducers couple relatively easily with enzymes, and so such biosensors are already on the market. Other bioreceptor-transducer couples are more difficult to construct, and are therefore still under study. Table 2.1 gives an idea of the approximate state of biosensor research as a function of the different possibilities of coupling between various bioreceptors and transducers. Table 2.1 also demonstrates how it is possible to classify biosensors with respect to either the bioreceptor or the transducer employed.

Transducer \ Biosensor		Enzyme	Cell, micro-organism	Immuno-agent	Animal/ plant tissue	Chemo-receptor
Electro-chemical cell	amperometric	xxx	xx	xx	x	
	potentiometric	xxx	xx	xx	x	x
Semiconductor (ISFET)		xx		x		
Thermometric (thermistors)		x		x		
Photometric (optical fibers)		x		x		
Piezoelectric (quartz)		x		x		
Chemomechanic		x				

Table 2.1 *Approximate state of biosensor research [(x) fundamental research, (xx) research and development of prototypes, (xxx) research and commercial devices].*

Considering that molecular recognition generally uses well-defined reaction types and that the detection method may be extremely varied, it is logical that biosensors should be classified primarily as a function of the bioreceptor used. However, a laboratory that only ever works with enzymes, for example, could use a classification according to the transducer employed (electrochemical, thermometric, etc). In what follows, we have opted to use the classification by bioreceptor because this component determines the primary action of the biosensor.

3

Construction of biosensors

Biosensor construction involves combining two elements with different characteristics. This entails three steps. First a bioreceptor is chosen, then a transducer, and finally the biological component is fixed to the transducer.

3.1 Choice of bioreceptor

The bioreceptor plays the role of a molecular recognition device. In the presence of the substance under investigation, it must produce a physicochemical effect that is detectable by the transducer. This may involve one of a number of processes, such as biocatalysis, immunological coupling, or chemoreception.

3.1.1 Enzymes

The biocatalyst employed is usually an enzyme, of which a large number are commercially available, for example, glucose oxidase and urease. Enzymes may also be extracted from one or more biological sources for extremely specific needs, and can be used alone or together with their cofactors, such as NAD and NADP.

The use of commercialized enzymes has a number of advantages, such as batch reproducibility, known characteristics and lifetimes, and rapid availability. Commercialized enzymes thus play an important role in the biosensors presently on the market. The disadvantage of purified enzymes lies in the fact that they are not always very stable, and often require the presence of their cofactors for proper operation. In some systems, a single enzyme is inadequate for the entire transformation desired and several are required for the whole enzymatic sequence, hence multienzyme sensors. Optimal conditions differ between one enzyme and the next, and a compromise is often adopted to ensure the optimal operation of the biosensor.

3.1.2 Microorganisms

If a biosensor makes use of complex reaction schemes then it may be more convenient to use microorganisms or cells. These structural entities possess all the necessary enzymes and cofactors in an environment that has already been optimized by nature. In the appropriate culture, microorganisms can reproduce and compensate for any loss in enzymatic activity with time.

3.1.3 Tissue and organelles

Another possibility for extending the field of biosensors is to employ sections of animal or plant tissue as sources of enzymatic material. Tissue has the advantage of cohesion, and has a structure that is robust enough to be attached directly to the transducer without having to resort to protein immobilization techniques.

(a) Animal and plant tissue

Plants are useful sources of enzymatic material for analytical chemistry. Using the appropriate transducer, very stable biosensors can be produced because the enzymes remain in their natural environment. Similarly, animal tissue can be considered as a useful bioreceptor for the selective detection of L-amino acids without any notable interference from D-amino acids. Sensor selectivity can also be improved by incorporation of an antimicrobial agent to avoid bacterial contamination. For example, it is recommended that 0.02 % sodium azide is used in the glutamine electrode [17].

(b) Organelles

Other biocatalysts are found in organelles such as lysosomes, chloroplasts, mitochondria and microsomes. These biomaterials have many enzymatic systems that can be exploited to construct biosensors. For instance, liver microsomes have a monooxygenase system with cytochrome P 450, which catalyses the hydroxylation of a large number of pharmaceutical derivatives, fatty acids, and steroidal hormones. This oxidation reaction can help to couple the organelle directly to an amperometric transducer [18].

3.1.4 *Immunoreceptors*

In contrast to biocatalysis, immunological coupling involves neither liberation of reaction products nor consumption of substrate. There is an extremely weak variation in potential arising from the antigen-antibody reaction [19], which, in practice, is difficult to detect. The response of the transducer can be improved by provoking a chemical reaction. This can be done by labelling the antigen with an enzyme that catalyses the production of electrochemically detectable species. The specificity of the biosensor is then determined by the antibody and its sensitivity determined by the enzyme performing the chemical amplification. The antibody is immobilized on the transducer using the well-known technique EIA (enzyme immunoassay) [20]. Another possibility is to use ion carriers (ionophores) as mediators between the immunological and electrochemical processes [21]. The antigen or hapten corresponding to the antibody under study is chemically coupled to an ionophore, forming a conjugate which is detectable with an electrochemical transducer.

3.1.5 *Chemoreceptors*

Many cellular membranes can be excited by chemical stimuli which induce a conformational change in their chemoreceptors. These changes, and the functional modifications that result, are reversible and provide an interesting model for the construction of biosensors [16]. Neuroreceptors can also be used in the detection of drugs, toxins, and certain other substances [22].

3.2 Choice of transducer

The transducer provides the evidence that the reaction of the bioreceptor has occurred. The choice of transducer depends on the reaction type, and the substances liberated or consumed (Table 2.1). Generally, the choice is the appropriate commercialized transducer for the detection method required. There are a large number of electrodes available for electrochemical detection, for example, electrodes that are sensitive to pH, to anions or cations, and to gases (pO_2, pCO_2 and pNH_3). The sensitive component may be bought independently (for example, a thermistor or an optical fiber), and then a bioreceptor attached to it.

The choice of transducer also depends on the intended application of the biosensor. If it is to be used in a biological environment, then it must fulfil the criteria of biocompatibility, especially with regard to the deposition of proteins, lipids, or cells on its surface. If it is to be used *in vivo*, then it should be of reduced size, and its form adapted so as not to cause excessive damage to living tissues. Furthermore, the possibility of the release of toxic, metallic, or polymeric components from the transducer should be taken into account when long-term implantation of the biosensor is intended.

The problem of chemical interference affects the choice of transducer. It is possible to couple urease with a number of different electrodes, such as pH, pNH_4, pCO_2, or pNH_3 electrodes. The best transducers for the determination of urea in a biological sample are the pCO_2 or pNH_3 electrodes because they have a hydrophobic membrane to eliminate all possible interference from cations and anions.

The transducer must also make optimal use of the physicochemical modification resulting from the biological reaction. If this reaction gives rise to a variation in enthalpy, then a very weak increase in local temperature will occur (of the order of 1/100th of a degree). Here a thermistor makes a better thermal transducer than a thermocouple.

Finally, the same transducer may be used in a number of different operation modes, which are chosen according to the objective desired. For example, an optical fiber can be used either as an extrinsic transducer, with a bioreactive component immobilized at its tip, or as an intrinsic transducer, exploiting the interaction between the immunoproteins at its surface and the evanescent surface waves.

3.3 Immobilization of bioreceptors

Biologically active proteins require direct or indirect immobilization on transducers to ensure maximal contact and response. Animal and plant tissue already have a membrane structure, and are an exception to this rule. Immobilization has the advantage of stabilizing the protein so that it can be used repetitively.

3.3.1 Immobilization of enzymes

(a) Physical entrapment

Enzymes are proteins with high molecular weights and sizes large enough to be trapped either in polyacrylamide gels or, more simply, by dialysis membranes. The first enzyme electrode was constructed by Updike and Hicks who trapped glucose oxidase in a layer of polyacrylamide gel [2], which they attached to the plastic membrane of an oxygen electrode. Acrylamide and N-N' methylene bis-acrylamide were copolymerized in the presence of the enzyme which becomes trapped in a polyacrylamide matrix. Although this copolymerization is inhibited by oxygen, it occurs in less than 10 minutes in the presence of light.

It is also possible to trap an enzyme to the sensitive component of an electrode using a dialysis membrane that prevents protein diffusion. Guibault and Shu [23] constructed a urea-sensitive enzyme electrode by spreading a suspension of urease over the surface of an electrode equipped with a nylon net, and covering the whole assembly with a dialysis membrane; the enzyme electrode is then rinsed and is ready for use.

Physical entrapment is now rarely used because the presence of the enzyme in solution does not give good long-term enzymatic activity. Chemical immobilization, using covalent bonds, ensures a greater long-term enzymatic stability [24], and a number of examples of this will now be described.

(b) Immobilization by cross-linking

Cross-linking is a process that uses a bi- or multifunctional agent to form a bridge between different biocatalytic species or proteins. The process results in a considerable increase in molecular weight, and the compounds or aggregates formed are insoluble.

It is possible to cross-link molecules of the same enzyme, or to coreticulate two or more different proteins (enzyme with enzyme, enzyme with protein, or more than one enzyme with a load protein such as BSA, bovine serum albumin). Even organelles or entire cells can be coreticulated. Coreticulation is a very important process because it makes multienzymatic immobilization accessible. The use of a load protein like albumin improves enzymatic activity because of the better mass distribution of the various proteins, but it does not alter the mechanical properties of the membranes produced.

Glutaraldehyde is the usual choice for a cross-linking agent [3]. This bifunctional agent has two aldehyde groups at its extremities which react with the amine group on the enzyme or protein, forming derivatives that are analogues of Schiff bases.

$$\text{(E)}{-}NH_2 + OCH\text{-}(CH_2)_3\text{-}CHO + H_2N{-}\text{(P)}$$
$$\downarrow$$
$$\text{(E)}{-}N{=}CH\text{-}(CH_2)_3\text{-}CH{=}N{-}\text{(P)}$$

It is also possible to use other bifunctional agents such as hexamethylene diisocyanate:

$$\text{(E)}{-}NH_2 + O{=}C{=}N\text{-}(CH_2)_6\text{-}N{=}C{=}O + H_2N{-}\text{(P)}$$
$$\downarrow$$
$$\text{(E)}{-}NH\text{-}CO\text{-}NH\text{-}(CH_2)_6\text{-}NH\text{-}CO\text{-}NH{-}\text{(P)}$$

Immersion method — The immersion method is used to immobilize urease [23] on a glass electrode which is sensitive to monovalent cations (Figure 3.1). The electrode is first immersed in a mixture of the enzyme, the albumin and the cross-linking agent, so that the bulb of the electrode is completely covered (a). The electrode is then rotated around its axis to give a homogeneous enzymatic layer (b). An O-ring joint may be incorporated at this stage (c). Finally, the electrode is rinsed with a glycine solution to elute and neutralize any excess cross-linking agent (d). The immersion method is simple, and suitable for immobilizing enzymes onto most transducers; it is particularly useful for small transducers [25].

Direct binding method — If an enzyme is expensive then it is important to use only the amount necessary for detection. Immobilization via the direct binding method (Figure 3.2) involves placing as little as 10 µl of enzymatic solution onto the tip of an electrode (a) and then an analogous quantity of the cross-linking agent (b) [26]. The enzyme can also be directly immobilized onto the hydrophobic membranes of gas-sensitive electrodes. In order to attach an enzyme to a fluorocarbon or polytetrafluoroethylene hydrophobic membrane, the enzyme is perfluoroalkylated [27] and then it absorbs directly onto the membrane.

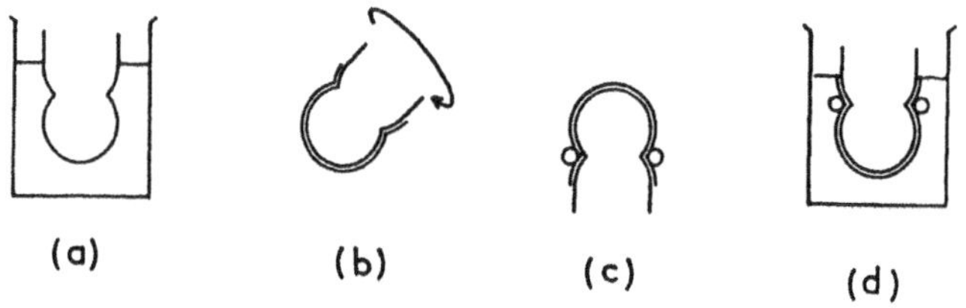

Figure 3.1 *Immobilization of enzymes by the immersion method.*

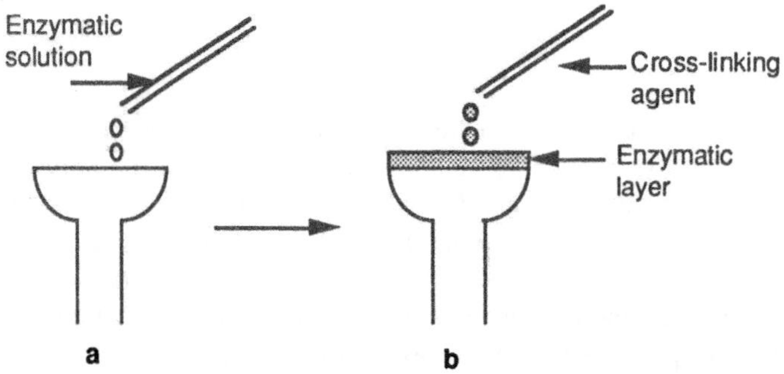

Figure 3.2 *Immobilization of enzymes by the direct binding method.*

Use of aerosols — A large number of applications require biosensors with a rapid response. Examples include alarm systems, in which urgency is vital, and flow injection analysis (FIA), in which a rapid response can increase the sample throughput and reduce the cost of analysis. Biosensor response times are very dependent on the thickness of the active layer (see § 4.2.1). Aerosol vaporization of dissolved compounds deposits films that are thin and homogeneous.

This technique was recently applied to the direct binding of enzymes onto transducers by vaporizing a cross-linking agent onto a protein that had been previously adsorbed onto the sensitive components of a transducer [28].

The sensitive tip of an electrode is washed and immersed in an enzymatic solution for about 20 minutes to ensure the adsorption of the enzyme on the glass (Figure 3.3). The electrode is then dried at 4 °C and set to rotate. Glutaraldehyde is vaporized onto the sensitive component from a distance large enough to prevent formation of droplets. After cross-linking the enzyme electrode is rinsed with water, and is then ready for use.

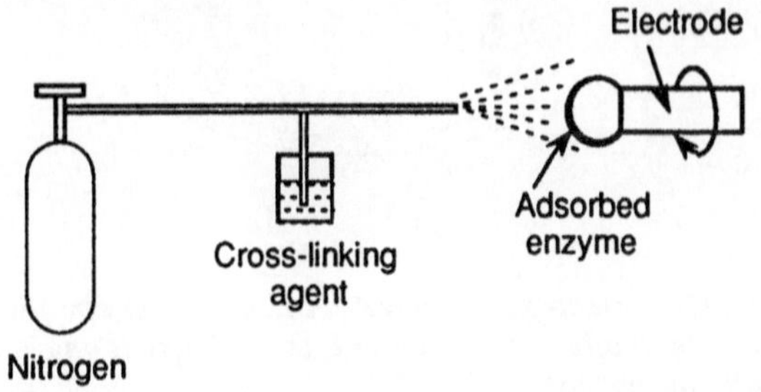

Figure 3.3 *Deposition of thin enzymatic membranes in the construction of glass enzyme electrodes.*

Films obtained in this way are extremely thin (1-2 μm) and the resulting biosensors have very short response times (5-10 seconds). This procedure has been used to immobilize penicillinase, urease, and acetylcholinesterase, and is also applicable for the immobilization of other proteins on a variety of supports [29].

Use of membranes — Biological compounds have limited lifetimes and so removable bioreceptors are useful. Hence the idea of immobilizing an enzyme on a membrane, and attaching the membrane to the transducer. The step also avoids the constraints of direct enzymatic immobilization on the transducer, and allows mass production, which improves the reproducibility of the biosensor signal.

Soft membranes can be obtained by coreticulation of the enzyme with a load protein like albumin or collagen. Such membranes have the same flexibility as the silicone or teflon membranes used in gas-diffusion electrodes. These two different membrane types can therefore be attached simultaneously to the same electrode (Figure 3.4), which is then the electrochemical transducer of the biosensor [30].

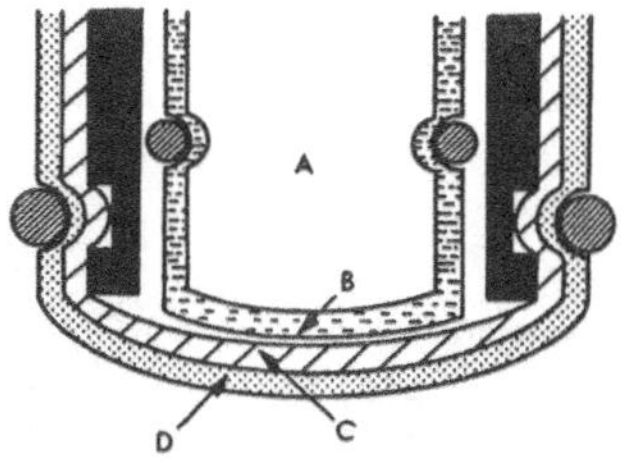

Figure 3.4 *pCO_2 electrode covered with a soft enzymatic membrane (A - glass electrode; B - electrolyte; C - silicone membrane; D - enzymatic membrane).*

Reinforced membranes have a fabric base, which improves the mechanical properties of enzymatic membranes [31] and facilitates their manipulation. The enzyme is immobilized on a large area of nylon net from which small disks of enzymatic membrane are cut in the exact dimensions of the transducer. These membranes come from the same batch and the same immobilization, and thus have reproducible activity. Membranes obtained in this way can be clipped onto the transducer [32, 33], which is much better than using screws which may twist or tear the membrane (Figure 3.5).

Prefunctionalized membranes use porous membranes, a number of which are commercially available. These contain ready-bound functional groups capable of reacting with the free sites of enzymes (for example, the membranes such as Immunodyne produced by Pall Industrie, or Ultra-Bind produced by Gelman Sciences). These prefunctionalized membranes are simply immersed in a solution containing the chosen enzyme to obtain an enzymatic membrane with useful mechanical properties [34].

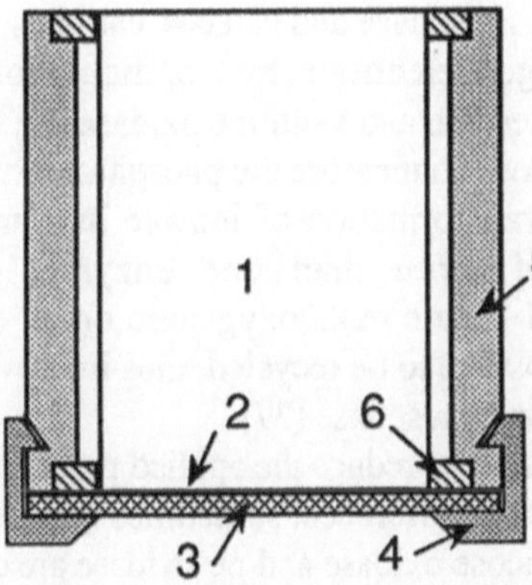

Figure 3.5 *Biosensor with a reinforced enzymatic membrane clipped onto a transducer (1 - transducer; 2 - sensitive component; 3 - enzymatic membrane; 4 - clip; 5 - sheath; 6 - washer).*

(c) Electromagnetic immobilization

If an enzyme can be incorporated into magnetic supports then it can be immobilized on a transducer using an electromagnetic field. For example, a biospecific interaction enables the fixation of glucose oxidase to magnetic beads carrying concanavalin A [35]. The magnetic particles containing the enzyme are then immobilized on the membrane of an oxygen electrode by applying a magnetic field produced by passing a continuous current through a solenoid. The current is regulated to give a homogeneous distribution of magnetic particles over the surface of the membrane. The distribution of the enzyme is homogeneous when the signal stops fluctuating because the magnetic interactions have stabilized. The electrode is then ready for use.

The advantage of this new method is that the biocatalytic component can be changed as often as desired. The enzyme is released by simply stopping the current in the solenoid. The only difficulty is the fixation of the enzyme to the magnetic particles; the biospecific interaction used is fragile and reversible, and the enzyme may be released under certain conditions of pH or ionic strength.

(d) Multienzymatic immobilization

The proteinic nature of enzymes means that they have a large number of reactive groups in common and can be simultaneously immobilized on a support. Glucose oxidase and catalase can thus be coimmobilized on the same oxygen electrode by coreticulation [3]. Similarly, phosphorylase nucleoside and xanthine oxidase can be coimmobilized on an oxygen electrode to measure the phosphate ions that are essential in the enzymatic transformation of inosine into uric acid [36]. The immobilization of more than one enzyme, such as malate dehydrogenase and lactate monooxygenase on an oxygen electrode, also enables the substrate to be recycled, thus improving the sensitivity of the enzyme electrode response [37].

The use of mediators to reduce the applied potential in amperometric electrodes, and hence interference, sometimes requires multienzymatic immobilization. Glucose oxidase and peroxidase are coimmobilized for the determination of glucose without interference from ascorbic and uric acids. The peroxidase catalyses the reaction of the hydrogen peroxide arising from the oxidation of glucose by the mediator hexacyanoferrate [38].

Multienzymatic immobilization also meets the needs of the catalysis of successive decomposition reactions which may require several different enzymes, but which will eventually lead to a product that is detectable by an electrode. This is the case for lactose determination using β-galactosidase (β-Gal) and glucose oxidase (GOD) [39]; the hydrogen peroxide formed is detectable by an amperometric platinum electrode.

$$\text{Lactose} \xrightarrow{\beta\text{-Gal}} \text{glucose} + \text{galactose}$$

$$\text{Glucose} + O_2 \xrightarrow{\text{GOD}} \text{gluconic acid} + H_2O_2$$

The coreticulation described above [3] is simple to perform during multienzymatic immobilization. The required enzymes are mixed with albumin, and enough glutaraldehyde is added to cross-link the various proteins. This method is also used for the coimmobilization of creatinin amidohydrolase, creatin amidinohydrolase and sarcosine oxidase, to construct a creatinin electrode, and for the coimmobilization of α-glucosidase and glucose oxidase for the determination of starch [40].

(e) Immobilization of cofactors

A cofactor (or coenzyme) may be immobilized on an insoluble support at the same time as an enzyme, provided that it forms an undissociable unity with the apoenzyme. For example, during the production of a glucose electrode, FAD (flavin adenine dinucleotide) is bound to glucose oxidase and does not require supplementary immobilization.

The problem of separate immobilization arises for free cofactors, such as NAD, NADP, ADP, and ATP, which pass from one apoenzyme to another during a reduction followed by a reoxidation. Biosensors that use dehydrogenase are of little use unless the cofactor is also immobilized. Moreover, the cofactor must remain sufficiently mobile after immobilization to pass from the active site of one apoenzyme to the active site of another for regeneration.

Physical immobilization of cofactors is impossible due to their small size. The cofactor must be covalently bound to either the enzyme or the insoluble support. This bond must be at the end of a spacer that is long enough to give the cofactor the required mobility (Figure 3.6).

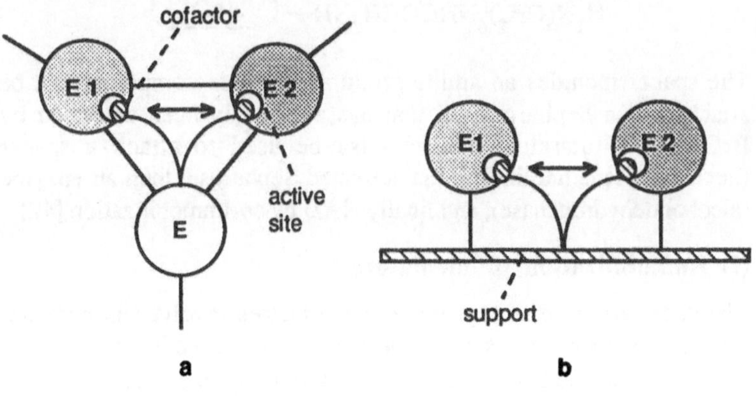

Figure 3.6 *Immobilization of cofactors (a) on the enzyme, or (b) on a support, using a spacer to give access to the active sites of both enzymes.*

Sometimes the regeneration of the cofactor does not require another enzyme, but simply another substrate. The cofactor is then immobilized at the active site of the enzyme [41] to obtain the enzymatic reaction with the first substrate, and then the cofactor (NAD) is regenerated by a second substrate (Figure 3.7).

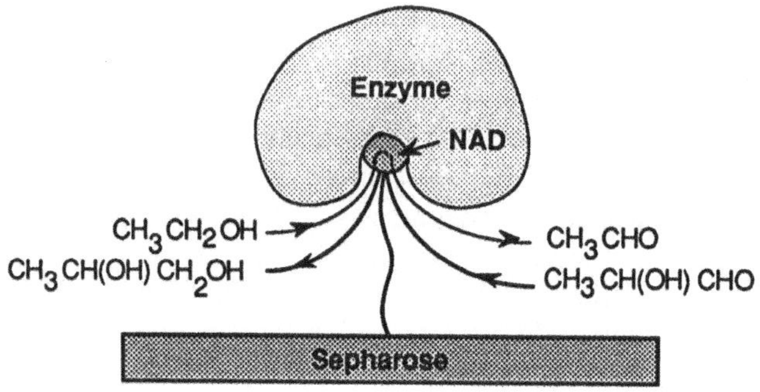

Figure 3.7 *Schematic representation of the alcohol dehydrogenase-NAD complex on Sepharose.*

Cofactors such as NAD can be immobilized on Sepharose gels by the transformation of NAD into carboxymethyl-NAD, and then, with the help of carbodiimide, into 8-(6-aminohexyl)carbamomethyl-NAD, which has the following structure:

$$H_2N(CH_2)_6NHCOCH_2NH\text{—}(\text{NAD})$$

The spacer includes an amine group and so this compound can be attached to a Sepharose gel that has previously been activated by BrCN [5]. Glutaraldehyde can also be used to attach a spacer (hexamethylenediamine) to the activated Sepharose, then an enzyme (alcohol dehydrogenase), and finally NAD by co-immobilization [42].

(f) Immobilization of mediators

Mediators are used to regenerate the cofactors involved in enzyme-catalysed redox reactions. Cofactor mobility is reduced if the enzyme is not in solution and a mediator is required to transport electrons between the electrode and the cofactor. The mediator must fulfil a number of other conditions:

1. The kinetics of their electron transfer must be both rapid and reversible, with respect to both the enzyme and the cofactor.
2. They must be stable in both the oxidized and reduced forms.
3. They must not react with the other species in solution (especially oxygen).

4. Their regeneration potential must be low and pH-independent.
5. They must be non-toxic for in-vivo applications.

Ferrocenes have characteristics that fulfil many of these criteria. They are hydrophobic enough for inclusion in carbon pastes and for adsorption onto graphite electrode surfaces. Unfortunately, their oxidized form, the ferricinium ion, is relatively soluble and may be released into solution.

Another approach is to use organic conducting salts, particularly those derived from N-methyl phenazine (NMP^+) and tetracyanoquinodimethane ($TCNQ^-$). These are compressed into pellets or attached to the electrode using a polymer [43]. The enzyme and its cofactor are then kept in the vicinity of the electrode using a dialysis membrane. Other conducting salts of $TCNQ^-$ are also used, such as tetrathiafulvalene (TTF^+), N-methyl acridium (NMA^+) and quinolinium (Q^+), all of which have good charge-transfer characteristics with respect to different types of electrodes. For example, TTF-TCNQ and PVC can be mixed into a paste with tetrahydrofuran (THF), and deposited on the platinum disc that forms the electrode (Figure 3.8). Immobilization of an enzyme is then performed simply by immersing the tip in an enzymatic solution [44]. Organic conducting salts are capable of regenerating both enzyme-bound cofactors (FAD) and free cofactors (NAD), and thus have many applications.

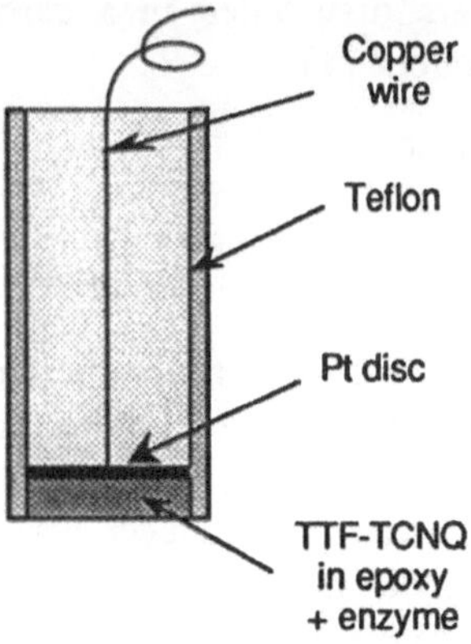

Figure 3.8 *Enzyme sensor using the organic conducting salt, TTF-TCNQ.*

Mediators are immobilized using either inert or electroactive polymer networks. Ion exchange polymers (for example, the fluorinated polymer Nafion) or conducting polymers, (polypyrroles, polyanilines, polyphenoxides and polyindoles) are also used [6, 45, 46]. These polymers are produced by electropolymerization, which facilitates their deposition on microtransducers and their application in field-effect transistors (FET). The enzyme can be attached to the surface of the electrode using electrochemical polymerization with or without the mediator. If the mediator is to be attached with the enzyme, then monomers containing the ferrocene group must be synthesized first; these monomers are ferrocenyl amidopropylpyrrole (FAPP) and ferrocenyl amidopentyl amidopropylpyrrole (FAPAPP). The polymer is then electrodeposited using a platinum electrode with a cyclic potential of between 0 and 1.0 V, and an aqueous solution of perchlorate containing varying quantities of pyrrole, FAPP and FAPAPP [45]. If the mediator is not to be attached with the enzyme, then polymerization of a solution of the pyrrole (or N-methylpyrrole) monomer and the enzyme is achieved by simply maintaining the potential of the electrode at that of the corresponding oxidation [46]. When sulfonated polyester polymer (type AQ from Eastman Kodak) is used, it is possible to dry the sensor thermally for 30 minutes at 50 °C which allows the immobilized enzyme to produce higher catalytic currents [47].

Finally, ferrocenes can be bound directly onto a polymer (for example, siloxane). The polymer is dissolved in a solvent (chloroform) and the ferrocenes are distributed in a carbon paste, and then immobilized on an electrode [48].

(g) Miniature sensors

The production of miniature biosensors meets the increasing demand for implantable microsensors for in-vivo clinical analysis.

Potentiometric sensors — Potentiometric biosensors are miniaturized by reducing the dimensions of the transducer on which the enzyme is deposited. Once the transducer is miniaturized, the enzyme is immobilized by immersing the transducer in a solution of the enzyme and a cross-linking agent, glutaraldehyde, as described above. An acetylcholinesterase electrode was constructed in this way for the determination of acetylcholine [25] using a glass microelectrode

(Figure 3.9).

Transducers based on pH-sensitive metal oxides can be used in the place of glass electrodes. One example is the pH electrode constructed by superficially oxidizing palladium either thermally at 800 °C or electrochemically. A potentiometric biosensor for urea is obtained by depositing urease onto the small sensitive component of this Pd/PdO electrode (Figure 3.10) [49]. Biosensors with transducers of the type Pd/PdO or Ir/IrO_2 are also used to determine inhibitors [50].

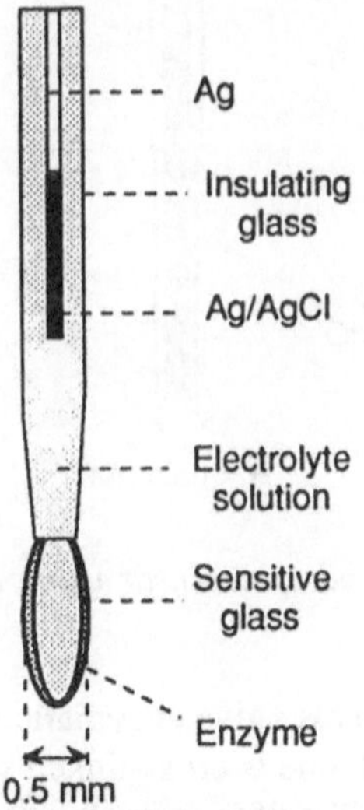

Figure 3.9 *pH micro-electrode used in the acetylcholine biosensor.*

Transducers with an ISFET (ion-sensitive field-effect transistor) structure can also be used as a basis for the construction of potentiometric biosensors. Their reduced size makes them particularly suitable for the construction of enzyme microsensors. The isolating oxide layer, SiO_2 is covered with an ultra-thin film of Si_3N_4 that is pH-sensitive but impermeable to ions. The enzyme is fixed to this film by modifying the Si_3N_4 surface using γ-aminopropyltriethoxy-silane (γ-APTES) and then covalently bonding the enzyme, for example, urease, to the surface with glutaraldehyde via the immersion method [51, 52].

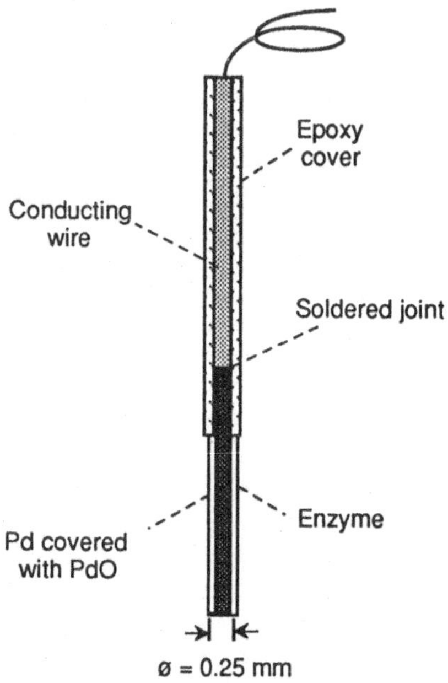

Figure 3.10 *Miniature urease biosensor using a Pd/PdO potentiometric transducer.*

Photolithography can also give enzymatic membranes for use with ISFET. Polyvinylpyrrolidone is cross-linked with a photo-reticulating agent, 2,5-bis-(4'-azido-2'-sulfonbenzal)cyclopentanone, which traps the enzyme in the polymer upon the action of ultraviolet light [53].

Amperometric sensors — The integrated circuit technology described above can also be used to construct miniature amperometric biosensors. A glucose-sensitive biosensor is obtained by attaching a glucose oxidase membrane to an oxygen microsensor. The enzymatic membrane is prepared from a triacetylcellulose membrane containing 1,8-diamino-4-aminomethyloctane and glutaraldehyde. The enzyme is bound to this membrane by immersion in an enzymatic solution [51]. Glucose oxidase can also be fixed to vitreous carbon electrodes after electrochemical oxidation and activation by the carbodiimide [54].

The miniaturization of amperometric electrodes is facilitated by their simple configurations, notably those of metallic electrodes. Platinum wire is especially suitable for the construction of a miniature transducer for a hydrogen peroxide sensor. The enzyme can then be immobilized on the end of the platinum anode. An enzyme electrode for glucose was constructed in this way (Figure 3.11). The anode is separated from the silver cathode by a glass capillary tube and the whole assembly has the form of a needle [55].

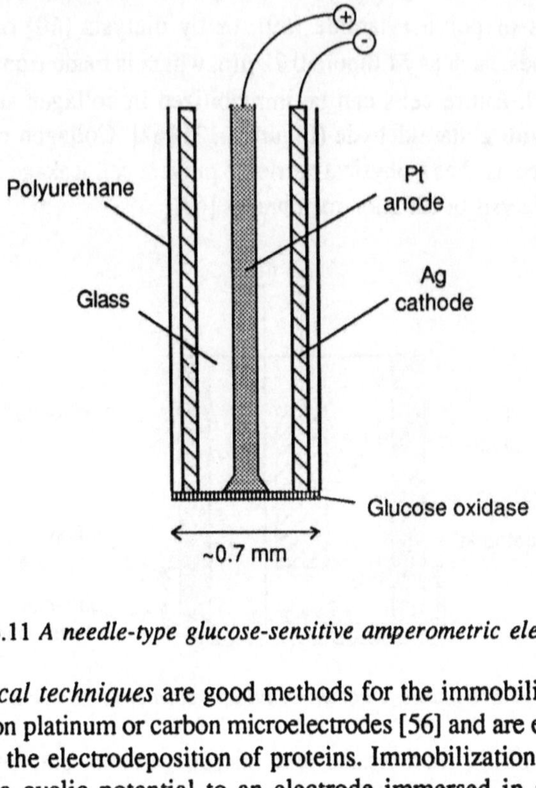

Figure 3.11 *A needle-type glucose-sensitive amperometric electrode.*

Electrical techniques are good methods for the immobilization of enzymes on platinum or carbon microelectrodes [56] and are especially useful for the electrodeposition of proteins. Immobilization involves applying a cyclic potential to an electrode immersed in a protein solution (BSA) and leaving the enzyme to impregnate the BSA [57]. Enzymes such as glucose oxidase can even be incorporated into a conducting polymer, such as ferrocene-containing polypyrrole, which can then be deposited electrochemically onto a platinum electrode [45]. Platinum black may also replace the conducting polymer. The enzyme is incorporated into the fine particles of platinum black during the electrodeposition. This technique lines the surface of the electrode homogeneously, whatever its original geometry, and biosensors with diameters as small as 10 μm can be constructed [58].

3.3.2 *Immobilization of microorganisms*

Microorganisms are much larger than enzymes and can thus be immobilized intact using physical entrapment. They may be trapped by agar gels or polyacrylamide [59], or by dialysis [60] or filtration membranes, such as Millipore 0.22 μm, which is made from cellulosic ester [61]. Entire cells can be immobilized in collagen suspensions treated with glutaraldehyde (Figure 3.12) [62]. Collagen membranes can also be used as a physical barrier to prevent cell leakage in a similar way to dialysis or filtration membranes [63].

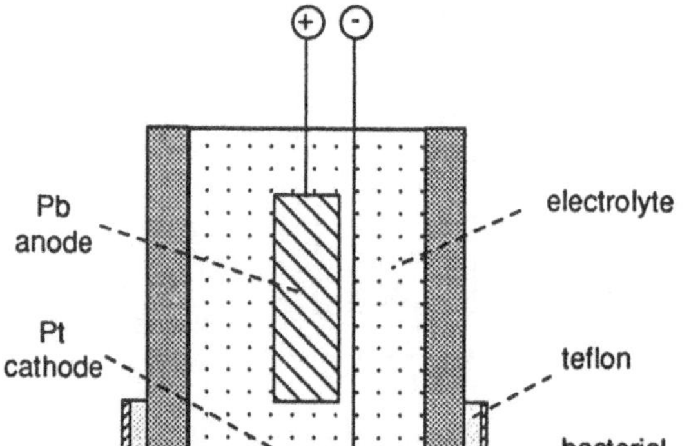

Figure 3.12 *Schematic diagram of an amperometric bacterial electrode.*

Hydrophilic membranes and microorganisms can be attached simultaneously to gas diffusion electrodes, such as those sensitive to pNH_3 [60, 61, 64, 65], pO_2 [62, 63, 66, 67] and pCO_2 [68]. The microorganisms are imprisoned in a space defined by a washer [69], or trapped by vacuum suction into the pores of a filtration cellulose acetate membrane [68, 70].

After culture of bacteria in a stirred solution, the cells are collected and washed. They are then placed on the electrode using a dialysis membrane, and rinsed in a buffer solution to eliminate any nutritive species which could interfere with the determination. The microbial sensor is preferably stored at 4 °C [64] in either a culture medium [64] or a phosphate buffer solution [70]. In a hybrid biosensor, an enzyme and a microorganism are immobilized on the same transducer [67]. The bacteria are immobilized by physical entrapment, and the enzyme is fixed by coreticulation which avoids any release of the protein.

Certain bacteria have the characteristic of forming thin sheets of cellulose in unstirred media. This is the case for acetic bacteria of the type *Acetobacter xylinum*. If these sheets are treated with glutaraldehyde then the membranes obtained have mechanical properties that allow their direct fixation to an oxygen electrode without the need for dialysis membranes [14].

3.3.3 Immobilization of immunoagents

A large variety of immunological sensors can be constructed by immobilizing immunoagents on different transducers. Immunosensors are classified into two categories: those which directly exploit a variation in parameters resulting from an immunological coupling; and those which have to resort to enzymes to liberate a product, or consume a cosubstrate, that is detectable by a transducer. The first step in both cases is to immobilize an antigen, an antibody, or a compound that has a bioaffinity with the analyte, on the transducer.

(a) Immobilization of antibodies

The aim of immobilizing an antibody on a transducer is usually to detect the corresponding antigen. Antibodies have a protein structure and can be immobilized in similar ways to enzymes, either directly or using membranes.

Direct immobilization on the transducer — Antibodies can be bound to a carbon electrode using covalent bonds. Carboxylic groups are generated on the electrode by electrochemical oxidation in a nitric acid and potassium bichromate medium [71, 72]. The antibody is attached to these groups using a cross-linking agent, such as carbodiimide. Another method physically adsorbs antibodies on metal films (gold or silver) deposited on glass. Surface plasmon resonance detection methods can then be employed [73].

Use of membranes — Antibodies can be immobilized on membranes, which can be easily attached to the sensitive component of a transducer. Bromo-acetylcellulose membranes are first immersed in a solution containing hexamethylene diamine, and then into diepoxy-butadiene, which produces the groups that will react with the antibody (Figure 3.13). Finally, the membrane with the bound antibodies is rinsed with a solution of ethanolamine to block any unreacted epoxy groups [74-76].

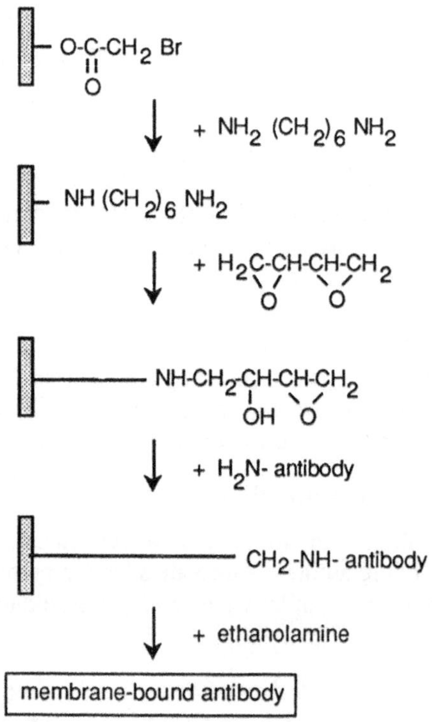

Figure 3.13 *Immobilization of an antibody on a bromo-acetylcellulose membrane.*

Antibodies can also be immobilized in nylon nets [77] or physically trapped in collagen membranes [78]. One example is the polyvinylbutyral membrane which is cast over the gate of a field-effect transistor (ISFET). The membrane is activated with glutaraldehyde before coupling with the antibody [79]. Antibodies can also be immobilized on an optical fiber using a removable sensing tip. If the antibody is in the liquid phase, then the antibody solution is trapped behind the sensing tip's 10 000 molecular weight cut-off membrane. If the antibody is in the solid phase, then it is immobilized on silica beads suspended in the sensing tip [80].

(b) Immobilization of antigens

Antigens do not always possess a protein structure and cannot always be immobilized in the same way as antibodies or enzymes. The antigenic determinant dinitrophenol (DNP) can be coupled with an ion carrier, such as dibenzo-18-crown-6 (DB18C6), and then immobilized in a polymeric membrane [81]. The crown ether is first converted into a trans-dinitro-derivative, and then into an arylamine derivative which is suitable for coupling with the hapten in the form of fluorodinitrobenzene. This hapten-ion carrier conjugate is dissolved in tetrahydrofuran, and dibutylsebacate is added. The conjugate is then incorporated into a polyvinylchloride (PVC) membrane. Finally, the antigenic membrane is cut up into small discs which are mounted on the tip of the electrode. The resulting biosensor is directly sensitive to antibodies [82].

Such biosensors use ion carriers which are directly connected to a conventional measuring system via the internal electrolyte solution. This is not essential because DNP has an affinity for potassium ions. A solid electrode with a graphite-loaded epoxy resin eliminates the need for a reference solution, and provides an excellent recovery of the base line potential [83]. This electrode is prepared by using the resin to fill the inside of a PVC tubing containing a platinum disc of the same diameter. A tetrahydrofuran solution of PVC, 10 mg/ml of the antigen DNP, and 15 % of trioctyl acetate is then placed in a hole drilled in the resin (Figure 3.14).

(c) Immobilization of compounds with bioaffinity

Compounds with bioaffinity can be used to prepare protein complexes by a process similar to immunological coupling. The compounds are immobilized on transducers using membranes obtained by the copolymerization of 1,8-diamino-4-amino-methyloctane with cellulose triacetate. The membranes are immersed in a solution containing ovalbumin and then HABA, the bioaffinity compound, is coupled with carbodiimide [84, 85].

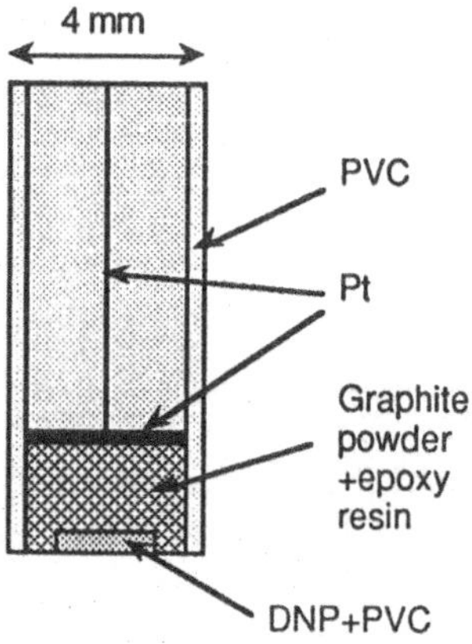

Figure 3.14 *Antigen-based antibody electrode using a solid state PVC matrix with DNP.*

(d) Enzymatic labelling

Even when an immunoagent is immobilized on an electrode or a membrane, the immunological coupling that occurs may not necessarily be detectable by the transducer; it must often be complemented by an enzymatic reaction. The antibody (or antigen) conjugate is labelled with an enzyme using the EIA (enzyme immunoassay) or ELISA (enzyme linked immunosorbent assay) techniques. The enzyme is fixed to the immunoagent using a bifunctional reagent, such as glutaraldehyde. The enzyme must be chosen as a function of the detection mode, so that there is a product appearing, or a substrate disappearing, that is detectable by the transducer. When the transducer is a pO_2 electrode, the antigen is labelled with glucose oxidase [77] or catalase [74, 76], which use or consume oxygen, respectively. When the transducer is a pNH_3 electrode, the label is urease, and ammonia [86] is liberated when immunological coupling occurs.

3.3.4 *Immobilization of tissue, organelles and chemoreceptors*

(a) Organelles and animal and plant tissue

Organelles and tissue of animal or plant origin constitute good sources of biocatalytic materials and the extracted enzymes are often employed in the construction of enzyme sensors. Using these materials in their original state keeps the enzymes in their natural environment and increases the stability of the catalyst.

First, a slice of the animal or plant tissue is cut. This is inserted between two semi-permeable membranes which are then attached to the transducer. The transducer is generally a pO_2, pCO_2 or pNH_3 electrode which already has its own gas-permeable hydrophobic membrane [17]. The animal tissue may be rabbit or bovine liver, rabbit muscle or intestinal mucosa. Cell components can even be used, for example, mitochondria. These organelles make the biosensor more selective because they contain more specific enzymes. A glutamine-sensitive biosensor was thus constructed by immobilizing mitochondria from pig kidney cells on a dialysis membrane. Plant tissue can also be immobilized on the sensitive element of an electrode using mechanical retention. Tissue from sugar beet [87], yellow squash, corn kernels, and cucumber [88] and spinach [89] leaves have been immobilized in this way.

Biosensors that use animal or plant tissue generally have a much longer response time than their corresponding enzyme sensors [17]. A more rapid response is obtained by incorporating the tissue in a carbon paste to give a greater contact between the biocatalytic site and the sensitive component [90]. Pieces of tissue are cut and crushed in a mortar. This is mixed with a mineral oil and graphite powder to give a paste that is spread over the sensitive tip of an electrode.

(b) Chemoreceptors

A chemoreceptor can be attached in its natural state to a potentiometric electrode to form biosensors which are called "receptrodes". One such receptor is the antennule of the blue crab *Callinectes sapidus*, whose nerve fibers are used as a transducer for the transmission of the nervous impulsions caused by olfaction or gustation [91]. The sensitive nerves are found near the crab endopod and are dissected before being fixed directly to the platinum electrode. The measurement is made using an Ag/AgCl reference electrode immersed in a saline solution, and the sensor is excited using different doses of stimuli injected into a carrier flow. These stimuli induce conformational changes in the receptor. The result is a change in the permeability of the excitable membrane, or a modification in the activity of enzymes bound to the membrane. The depolarization of the axonal nerve gives rise to a measurable potential [92].

Biosensors based on neuroreceptors can measure toxins or other chemical agents used in warfare. The use of acetylcholine receptors enables the determination of acetylcholine by the detection of a specific impedance compared with other neurotransmitters [93]. This receptor is associated with immobilized acetylcholinesterase, and is sensitive to its inhibitors, organophosphorus compounds and carbamates. The receptor is physically adsorbed onto a suitable sensor by immersing the sensitive tip of the sensor in a liposome solution containing the receptor.

4

Enzyme sensors

An enzyme sensor can be considered as the combination of a transducer and a thin enzymatic layer, which normally measures the concentration of a substrate. The enzymatic reaction transforms the substrate into a reaction product that is detectable by the transducer. An extension of this definition is that the concentration of any substance can be measured provided that its presence affects the rate of an enzymatic reaction; this is especially true for enzyme inhibitors. We will first examine the principle of enzyme sensors and the theoretical and practical aspects of their operation. We will then describe the various sensors, classified according to their detection mode.

4.1 Principles of operation

A schematic representation of an enzyme sensor is given in Figure 4.1. The sensitive surface of the transducer is in contact with an enzymatic layer, and it is assumed that there is no mass transfer across this interface. The external surface of the enzymatic layer is immersed in a solution containing the substrate under study. The substrate migrates towards the interior of the layer and is converted into reaction products when it reacts with the immobilized enzyme. In order to achieve rapid equilibrium of concentrations, the enzymatic membrane must be as thin as possible. The solution must also be well stirred to ensure a constant supply of substrate. In summary, the different steps during the operation of an enzyme sensor are:

1. transport of the substrate from the bulk of the solution towards the enzymatic layer,
2. diffusion of the substrate within this layer, accompanied by the

enzymatic transformation of the substrate into reaction product,
3. migration of the product towards the transducer,
4. conversion of the concentration of the product at this interface into an electrical signal by the transducer.

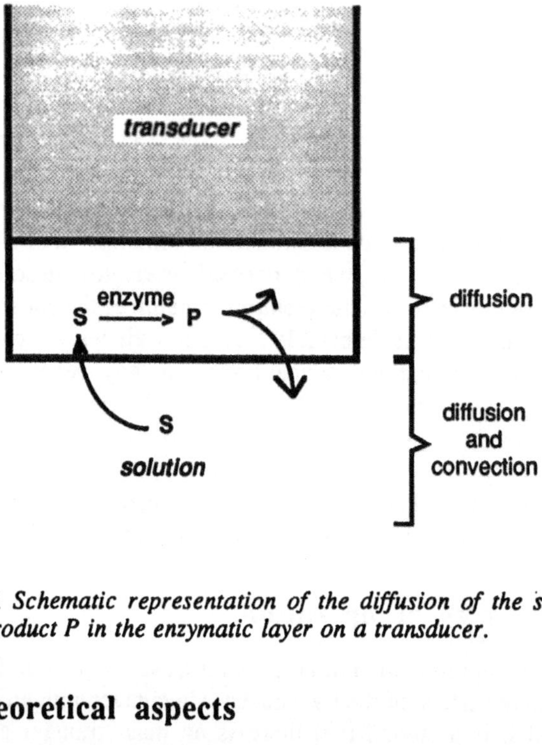

Figure 4.1 *Schematic representation of the diffusion of the substrate S and the product P in the enzymatic layer on a transducer.*

4.2 Theoretical aspects

The presence of the enzyme ensures the transformation of the substrate into the reaction product according to the following reaction:

$$E + S \underset{k_{-1}}{\overset{k_{+1}}{\rightleftarrows}} ES \overset{k_{+2}}{\longrightarrow} E + P \qquad (1)$$

where E represents the enzyme, S the substrate and P the reaction product; k_{+1}, k_{-1} and k_{+2} are the rate constants of the reactions. The reaction rate can then be written as:

$$V = \frac{d[P]}{dt} = - \frac{d[S]}{dt} = V_m \frac{[S]}{K_m + [S]} \quad (2)$$

where K_m is the Michaelis constant ($K_m = (k_{-1} + k_{+2})/k_{+1}$) and V_m is the maximal rate for $[S] >> K_m$ ($V_m = k_{+2} [E]_0$, where $[E]_0$ is the initial concentration of the enzyme). [S] and [P] are the concentrations of the substrate and product, respectively. For a given enzyme, the rate of the reaction V is a function of the ratio $[S]/K_m$, and:

when $[S]/K_m \longrightarrow 0$, $V \longrightarrow (V_m/K_m)[S]$ (first order kinetics with respect to [S]),

when $[S]/K_m$ becomes large, $V \longrightarrow V_m$ (zeroth order kinetics with respect to [S]).

Inside the enzymatic layer, the reaction rate is coupled with the rate of diffusion of the various substances present. This phenomenon is governed by the following equations:

$$\frac{\partial[S]}{\partial t} - D_S \frac{\partial^2[S]}{\partial x^2} + \frac{V_m [S]}{K_m + [S]} = 0 \quad (3)$$

$$\frac{\partial[P]}{\partial t} - D_P \frac{\partial^2[P]}{\partial x^2} - \frac{V_m [S]}{K_m + [S]} = 0 \quad (4)$$

where t is the reaction time, D_S and D_P are the diffusion coefficients of the substrate S and the product P at the interior of the active layer, and x is the distance between a given point in the active layer and the external surface of the sensor.

4.2.1 Response in the transient state

Equations (3) and (4) show how the concentrations of the substrate and product at the transducer depend on the enzyme Michaelis constant K_m, the activity V_m of the enzymatic layer, the thickness of the layer, and the diffusion coefficients of the substrate and product. The product concentration at the external surface of the electrode is assumed to be zero because the diffusion of the species occurs essentially in the active layer. The temperature, V_m, K_m, and the coefficients of diffusion of the substrate and the product, are all considered as constant throughout the entire layer. It is also assumed that there is no consumption of the substrate or the product by the transducer.

The following dimensionless reduced variables are used:

$$s = \frac{[S]}{K_m},\ p = \frac{[P]}{K_m},\ \sigma = \frac{V_m\, e^2}{K_m . D_S},\ t' = \frac{t}{e^2/D_S},\ x' = \frac{x}{e}$$

where e is the thickness of the layer. This leads to:

$$\frac{\partial s}{\partial t'} - \frac{\partial^2 s}{\partial x'^2} + \sigma \frac{s}{1+s} = 0 \qquad (3')$$

$$\frac{\partial p}{\partial t'} - \frac{D_P}{D_S}\frac{\partial^2 s}{\partial x'^2} - \sigma \frac{s}{1+s} = 0 \qquad (4')$$

Equations (3') and (4') involve dimensionless variables and are easily solved by a computer. These equations can be used to calculate the concentration profiles of the substrate and product in the transient state (Figure 4.2). Calculation shows that these profiles depend on a single dimensionless parameter:

$$\sigma = \frac{V_m}{K_m} \cdot \frac{e^2}{D_S} \qquad (5)$$

In Figure 4.2, the concentrations of the substrate S (broken line) and the product P (solid line) are expressed in terms of K_m. Time is expressed as e^2/D_S. At time $t = 0$, the substrate and product concentrations in the membrane are zero. The concentration of the substrate in the solution is always kept constant and equal to K_m. The following values are used for the enzymatic membrane:

$V_m = 1.92 \times 10^{-6}$ mol cm^{-3} s^{-1}
$K_m = 2.10 \times 10^{-5}$ mol cm^{-3}
$e = 2.5 \times 10^{-3}$ cm
$D_S = D_P = 1.2 \times 10^{-5}$ cm^2 s^{-1}.

The time taken to reach the steady state is then:

$t = 0.7\ e^2/D_S = 36$ seconds.

The ratio e^2/D_S determines the response time of the biosensor because it expresses the time taken for the enzymatic layer to reach the steady state. This response time can be reduced by modifying the membrane so that it is extremely permeable to the substrate (increase D_S) or very thin (decrease e). Reduction in enzymatic membrane thickness is the most effective method because this value is squared in the expression for the response time. In reality, an excessive decrease in the thickness of the membrane will also affect its mechanical properties. Moreover, the response time of the biosensor can never be less than the response time of the transducer.

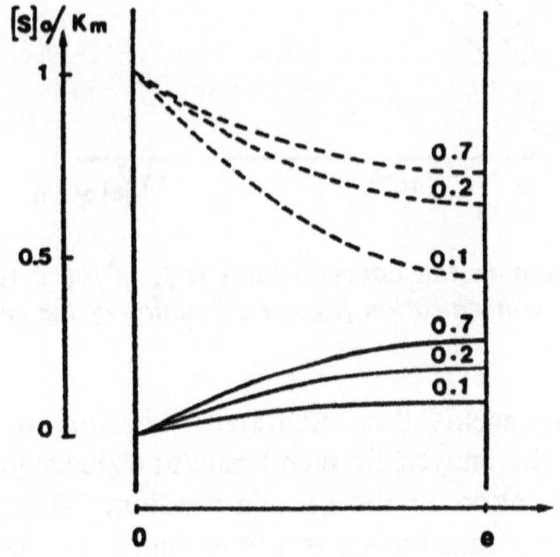

Figure 4.2 *The transient-state concentration profiles of the substrate (---) and product (—) in an enzymatic layer on a potentiometric transducer.*

4.2.2 Response in the steady state

The response of an enzyme sensor in the steady state depends largely on the ratio of the substrate concentration [S] to the enzyme Michaelis constant K_m (Figure 4.3). When $[S]/K_m$ is large, the reaction rate reaches a maximal value V_m, which is proportional to the number of active sites of the immobilized enzyme. The reaction rate is then independent of the substrate concentration, and the product concentration at the contact with the transducer is the same for all high substrate concentrations.

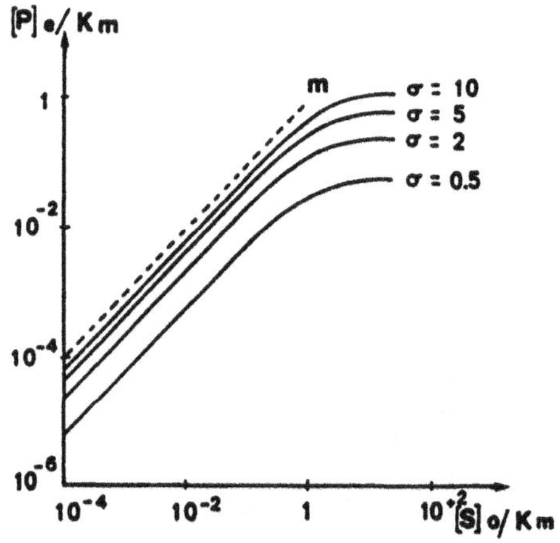

Figure 4.3 *Variation in the concentration* $[P]_e$ *of the detected product with the substrate concentration* $[S]_0$ *as a function of the enzyme loading factor* σ.

Figure 4.3 represents the calculated variation in the product concentration at the enzymatic membrane/transducer interface with substrate concentration in the sample medium. These curves are presented in log-log coordinates as a function of the enzyme loading factor in the membrane, σ (Equation (5)) [94]. If the transducer response is proportional to the logarithm of the product concentration (for example, in potentiometric sensors) then these curves are actually the theoretical response curves of the biosensor. Their form is analogous to the curves for enzymatic kinetics in the homogeneous phase, especially with regard to the saturation of the response at high substrate concentrations ($[S] >> K_m$). The curves in Figure 4.3 also have the following characteristics:

1. They are linear at substrate concentrations lower than about 0.1 K_m, but the linear zone increases with σ.

2. Their slope does not depend on the value of σ.
3. When σ is large, they approach the theoretical straight line corresponding to the point where the concentrations are equal, $[P]_e = [S]_0$.

The most important result of this theoretical simulation is that the quantity of enzyme in the membrane determines the linear zone in the response to the substrate concentration. This zone corresponds to first order enzymatic kinetics with respect to substrate concentration, whereas the region with a plateau has zeroth order kinetics.

(a) First order kinetics

First order kinetics occur when $[S] << K_m$. Under these conditions, Equations (3) and (4) achieve a steady state where $d[S]/dt = 0$:

$$D_S \frac{\partial^2[S]}{\partial x^2} = \frac{V_m}{K_m}[S] \quad (6)$$

$$D_P \frac{\partial^2[P]}{\partial x^2} = -\frac{V_m}{K_m}[S] \quad (7)$$

The product concentration detected by the transducer is obtained by integration [95]:

$$[P]_e = [S]_0 \, [1 - 1/\cosh(\lambda e)] \quad (8)$$

where $\lambda^2 = V_m/(D_S K_m)$, and $\cosh(\lambda e)$ is the hyperbolic cosine of (λe).

The detectable product concentration $[P]_e$ is proportional to the concentration of the substrate $[S]_0$, provided that the latter is small compared with the enzyme Michaelis constant. This confirms the results of the digital simulation in log-log coordinates, and from Equation (8) we obtain:

$$\log [P]_e = \log [S]_0 + \log [1 - 1/\cosh(\lambda e)] \quad (9)$$

(b) Zeroth order kinetics

Zeroth order kinetics occur when $[S] >> K_m$ in the steady state, and Equations (3) and (4) become:

$$D_S \frac{\partial^2[S]}{\partial x^2} = V_m \qquad (10)$$

$$D_P \frac{\partial^2[P]}{\partial x^2} = -V_m \qquad (11)$$

Integration then gives [95]:

$$[P]_e = V_m \frac{e^2}{2D_P} \qquad (12)$$

The product concentration detected by the transducer is independent of both the substrate concentration and K_m. When the substrate concentration is very high ($[S] >> K_m$), the biosensor is no longer capable of determining the substrate but may determine inhibitors which affect the maximal rate of the enzymatic reaction V_m [56].

4.2.3 Stability of enzyme sensors

The enzyme is the essential component of an enzyme sensor. Its protein structure is sensitive to denaturation by temperature, pH or simply by aging. This gradual loss in activity generates a decrease in V_m and there is a decrease in the value of σ because the other determining values remain unchanged (K_m, e and D_S) as can be seen in Equation (5). Figure 4.3 shows that this reduction in σ corresponds to a downward displacement of the curves, or to a decrease in the electrical signal from a transducer sensitive to the reaction product.

An examination of Figure 4.3 indicates that this loss in signal accelerates at very low values of σ, when the enzymatic membrane is not very active at all. However, if the membrane contains a large amount of enzyme (and σ is high) then the response curves become closer and closer, and begin to merge with the straight line corresponding to the situation when the concentrations are equal $[P]_e = [S]_0$. This is especially true when $[S]_0 < 0.1\ K_m$, which is the operating region of the sensor.

The stability of enzyme sensors is thus related to the enzyme loading factor σ of the active membrane. When the membrane contains more active enzyme, the biosensor is more stable and has a longer lifetime. We will see, however, that it is not always possible to satisfy this condition, notably in the case of inhibitor determination where a compromise must be adopted.

4.2.4 *Study of enzymatic inhibition*

Inside living cells, enzymes exist in an extremely structured environment, which may have the form of a gel, a membrane or other solid state assemblies. Under these conditions, the behavior of immobilized enzymes towards their inhibitors is very different from that in the homogeneous phase. The mode of enzymatic inhibition is variable from one inhibitor to another, and may also be reversible or irreversible. Inhibition may be observed with enzyme sensors without having to resort to techniques that require long extractions, such as dialysis or gel filtration, which are normally used to separate an inhibitor from a soluble enzyme. The response of the enzyme sensor provides a continuous measure of the activity of the immobilized enzyme, whether or not its inhibitor is present. Enzyme sensors thus allow convenient and rapid study of the inhibition of enzymes and their subsequent reactivation.

(a) Different types of inhibition

Reversible inhibition is characterized by an equilibrium between the enzyme and the inhibitor, defined by an equilibrium constant, K_I, which represents the affinity of the inhibitor for its enzyme. Reversible inhibitors are classed into three groups according to their effect on the enzymatic site (Figure 4.4). The inhibition is "competitive" when the inhibitor and the substrate are in competition for the same site, and the degree of inhibition decreases as the substrate concentration increases. When the degree of inhibition increases with the substrate concentration, the inhibition is called "uncompetitive". Finally, if the degree of inhibition is independent of the substrate concentration, then the inhibition is said to be "non-competitive".

Competitive inhibition — Competitive inhibition is represented by:

$$
\begin{array}{l}
E + S \rightleftarrows ES \longrightarrow E + P \\
+ \\
I \\
(K_I) \uparrow\downarrow \\
EI
\end{array}
$$

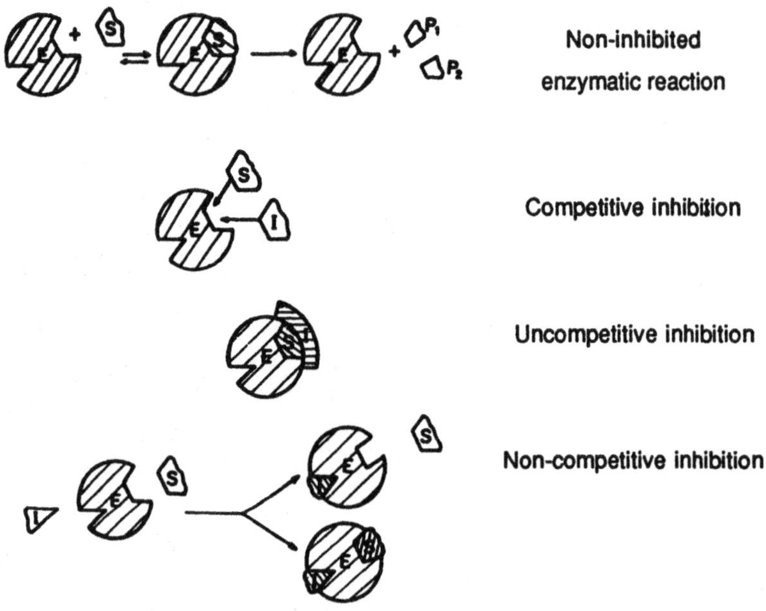

Figure 4.4 *Schematic representation of the principal modes of inhibition.*

where K_I is the dissociation constant of the complex EI. The reaction rate in the presence of the inhibitor, V_I, can be written using the steady-state approximation:

$$V_I = V_m \frac{[S]}{K_m \left(1 + \frac{[I]}{K_I}\right) + [S]}$$

$$\text{if } i = \frac{[I]}{K_I} \text{ and } s = \frac{[S]}{K_m}$$

$$V_I = V_m \frac{s}{1 + i + s}$$

The degree of inhibition, ρ, is related to the rate of the non-inhibited enzymatic reaction V (Equation (2)) and is given by:

$$\rho = 1 - \frac{V_I}{V} \quad \text{with} \quad 0 < \rho < 1,$$

$$\text{and because } V = V_m \frac{s}{1 + s}$$

$$\text{then } \rho = \frac{i}{1 + i + s}$$

Thus, the degree of inhibition diminishes as the substrate concentration increases. Competitive inhibition is therefore completely removed in the presence of an excess of substrate.

Uncompetitive inhibition — Uncompetitive inhibitors only interact with ES complexes, according to the following reaction scheme:

$$E + S \rightleftarrows ES \longrightarrow E + P$$
$$+$$
$$I$$
$$(K_I) \downarrow\uparrow$$
$$ESI$$

Using the steady-state approximation, we obtain:

$$V_I = \frac{V_m}{1 + \frac{[I]}{K_I}} \cdot \frac{[S]}{1 + \frac{K_m}{1 + \frac{[I]}{K_I}}}$$

$$V_I = V_m \frac{s}{1 + s + si}$$

and hence the degree of inhibition:

$$\rho = \frac{si}{1 + s + si}$$

Thus, the degree of inhibition, ρ, increases with the substrate concentration, s, for the same inhibitor concentration, i.

Non-competitive inhibition — Non-competitive inhibitors bind equally well to both the enzyme E and the complex ES. This is because the inhibitor interacts with a site on the enzyme which is not its active site.

$$\begin{array}{ccccc} & E + S & \rightleftarrows & ES & \longrightarrow E + P \\ & + & & + & \\ & I & & I & \\ (K_I) & \downarrow\uparrow & & \downarrow\uparrow & (K'_I) \\ & EI + S & \rightleftarrows & ESI & \end{array}$$

This results in the existence of three complexes ES, EI and ESI, of which only ES can yield the product of the enzymatic reaction. If it is assumed that the interaction between the enzyme and inhibitor is not perturbed by the presence of the substrate on the enzyme ($K_I \simeq K'_I$) then the rate of the inhibited enzymatic reaction can be written as:

$$V_I = \frac{V_m}{1 + \frac{[I]}{K_I}} \cdot \frac{[S]}{K_m + [S]}$$

The degree of inhibition obtained, $\rho = i/(1 + i)$, is independent of the substrate concentration. This type of inhibition is therefore not removed by an excess of substrate.

In summary, it can be considered that the rate of the non-inhibited enzymatic reaction is modified in the presence of an inhibitor by a term r, which affects either V_m or K_m, or both.

$$r = 1 + \frac{[I]}{K_I}$$

The new values of V_p or K_p for an inhibited reaction allow the characterization of the inhibition mode present. These values may be compared to the values V_m or K_m for the non-inhibited enzymatic reaction.

$$K_p = r\ K_m \quad \text{and} \quad V_p = V_m \qquad \text{(competitive inhibition)}$$

$$K_p = \frac{K_m}{r} \quad \text{and} \quad V_p = \frac{V_m}{r} \qquad \text{(uncompetitive inhibition)}$$

$$K_p = K_m \quad \text{and} \quad V_p = \frac{V_m}{r} \qquad \text{(non-competitive inhibition)}$$

These results are relevant when the enzyme, the substrate and the inhibitor are all dissolved in homogeneous solution.

(b) Monitoring enzymatic inhibition

When the enzyme is immobilized on a solid support, the reaction occurs in the heterogeneous phase because the active sites are in the solid phase and the substrate is in the aqueous phase. The rates of enzymatic reactions, V_I, described above for the various inhibitors, are insufficient for the characterization of such a substrate transformation due to constraints connected with the micro-environment and the diffusion of the species.

The concentration of the substrate varies from one point in the solid support to another, and so the rate of the enzymatic reaction is not the same as that occurring at the solid surface. During the reaction, a profile of the substrate and product concentrations is established across the support such that the substrate concentration in the active layer is always lower than that in solution. The saturation of the immobilized enzyme therefore requires higher substrate concentrations in the bulk solution than those normally obtained for the enzyme in solution [96]. This means that it is no longer possible to obtain the values V_p and K_p for the enzyme in the presence of its inhibitor from the Lineweaver-Burk linear transformations [97] because they are only applicable in the homogeneous phase.

A biosensor that uses immobilized enzymes thus appears to be a convenient and effective device for gaining an insight into the activity of enzymes that act at the interior of cells and organisms.

When the enzyme sensor is in contact with an inhibitor, the reaction rate diminishes according to the inhibition mode involved. The substrate and product concentrations in the steady state can be obtained by solving Equations (3) and (4), provided that V_m and K_m are replaced by V_p and K_p. Figure 4.5 shows the concentration profiles of the substrate (broken line) and the product (solid line) in terms of K_m.

These concentrations are simulated for different inhibitor concentrations which are expressed in terms of $r = 1 + ([I]/K_I)$. The following values were used in this calculation:

$V_m = 10^{-5}$ mol cm^{-3}s^{-1}
$K_m = 10^{-5}$ mol cm^{-3}
$e = 2.5 \times 10^{-3}$ cm
$D_S = D_P = 1.2 \times 10^{-5}$ cm^2 s^{-1}

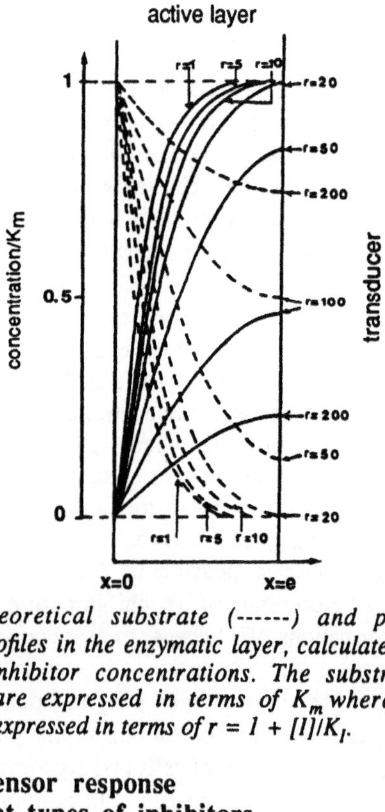

Figure 4.5 *Theoretical substrate (------) and product (———–) concentration profiles in the enzymatic layer, calculated as a function of uncompetitive inhibitor concentrations. The substrate and product concentrations are expressed in terms of K_m whereas the inhibitor concentration is expressed in terms of $r = 1 + [I]/K_I$.*

(c) Enzyme sensor response to different types of inhibitors

The digital simulation above gives the concentration of the substrate $[S]_e$ and, more importantly, that of the product $[P]_e$ at the $[P]_e$-sensitive base transducer. Using this information, the response of the enzyme sensor in the presence of different inhibitor concentrations expressed in terms of r, can be simulated using log-log coordinates (Figure 4.6). In this diagram, the maximal response of a competitive inhibitor (a) is independent of the inhibitor concentration. This region actually has zeroth order kinetics and the relationship

$$\log [P]_e = \log \frac{V_m e^2}{2D}$$

is independent of K_m. Competitive inhibitors only depend on the value of K_m, and do not affect the plateau in the maximal response curve. In contrast, the response of a potentiometric electrode in the linear region that corresponds to first order kinetics is given by:

$$\log [P]_e = \log [S]_0 + \log \left(1 - \frac{1}{\cosh(\lambda e)}\right)$$

$$\text{with} \quad (\lambda e)^2 = \sigma = \frac{V_m}{K_m} \frac{e^2}{D}$$

Competitive inhibitors increase the value of K_m, diminishing σ and thus generating a decrease of log $[P]_e$ in the response of the biosensor. This does not apply to uncompetitive inhibitors (b) which do not change the ratio V_m/K_m and thus the linear region of the response curves is the same. In contrast, such inhibitors lower the plateau in the response curve by reducing V_m to V_m/r. Non-competitive inhibitors (c) only affect the enzyme loading factor, σ, through V_m. Their presence in increasing concentrations displaces the response curves in the same way as a reduction in the value of the parameter σ (Figure 4.3).

Consequently, it is possible to determine the inhibition mode of an enzyme by a particular inhibitor by simply observing the displacement of the response curves of a *potentiometric* electrode on which the enzyme is immobilized. Inhibition is competitive if the plateau that corresponds to the maximal response is unchanged. When the response to low substrate concentrations is not modified in the presence of inhibitors then the inhibition is uncompetitive. Non-competitive inhibition gives rise to a response curve that has no point in common with the response curve in the absence of the inhibitor. These results for simulated response have been confirmed experimentally [56].

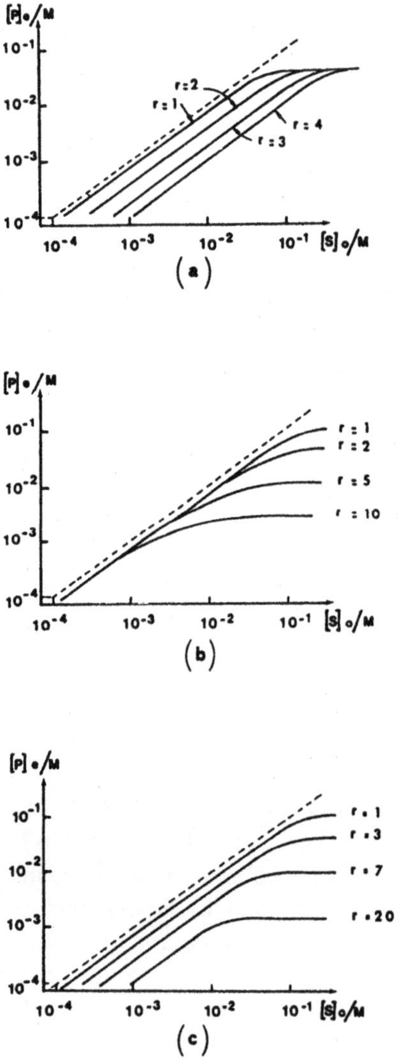

Figure 4.6 *Response of an enzyme electrode as a function of different inhibition modes: (a) competitive, (b) uncompetitive, (c) non-competitive. Inhibitor concentrations are expressed in terms of* $r = 1 + [I]/K_I$.

4.3 Practical aspects

Specific application of an enzyme sensor requires precise information about its response time, the range and the reproducibility of its calibration curve, its detection limit, and any other parameters that determine the stability of its response.

4.3.1 Response time

The response time of a biosensor is normally the time taken for it to reach a steady state where it is no longer possible to detect any variation in the signal. In practice, this is very difficult to evaluate precisely because of the asymptotic nature of the curve (Figure 4.7). For this reason, the t_{98} value corresponds to the time taken to reach 98 % of the sensor response and is often used instead of the response time.

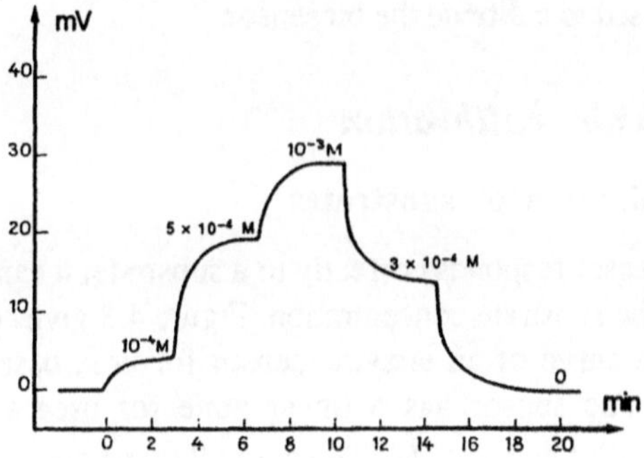

Figure 4.7 *Response of the AChE electrode as a function of time for different acetylcholine concentrations.*

In the theoretical section above, we studied the response time due to the enzymatic layer. This is the time taken to reach a state where the substrate and product concentrations are constant. In practice, the transducer cannot immediately transform a chemical signal into an electrical one, but requires a certain time, which depends on the transducer used. Thus, a urease membrane fixed onto a gas diffusion electrode (pCO_2) responds in 2 to 3 minutes [30], whereas the same membrane fixed to a monovalent cationic electrode responds in just under a minute [98, 99]. This difference arises from the fact that the pCO_2 electrode has a hydrophobic membrane which limits the diffusion of CO_2. A new technique for the immobilization of proteins recently combined adsorption and cross-linking processes to yield ultra-thin enzymatic films (1-2 μm thick). Electrodes covered with such films respond in a few seconds [28], as opposed to a few minutes, the time usually observed (Figure 4.7).

Enzyme sensors operate by consuming substrate and so stirring the sample medium reduces the response time of a biosensor since it favors the movement of the substrate towards the active membrane. A large membrane permeability also facilitates diffusion of all the species present, and a steady state is achieved much more rapidly.

Finally, if the response time of a biosensor is considered to be too long, it is possible to work with the *derivative of the response curve.* Each value on this curve corresponds to the slope of the response curve and can be used to calibrate the biosensor.

4.3.2 Sensor calibration

(a) Determination of substrates

Once a biosensor responds correctly to a substrate, it can be calibrated by varying the substrate concentration. Figure 4.8 gives an example of a calibration curve of an enzyme sensor for urea based on a pNH_4 transducer. The sensor has a linear zone for urea concentrations between 3×10^{-4} and 10^{-2} M. The curve flattens out on both sides of this linear zone. The upper plateau corresponds to high concentrations of urea ($[S] >> K_m$), a situation which saturates the active sites of the enzyme. The lower plateau corresponds to the *detection limit* of the transducer. The only useful part of the curve is the linear zone where a variation in substrate concentration gives a variation in the potential at the transducer. The slope of the curve in this zone corresponds to the *sensitivity* of the biosensor because it expresses a variation in the signal obtained as a function of the analyte concentration [100].

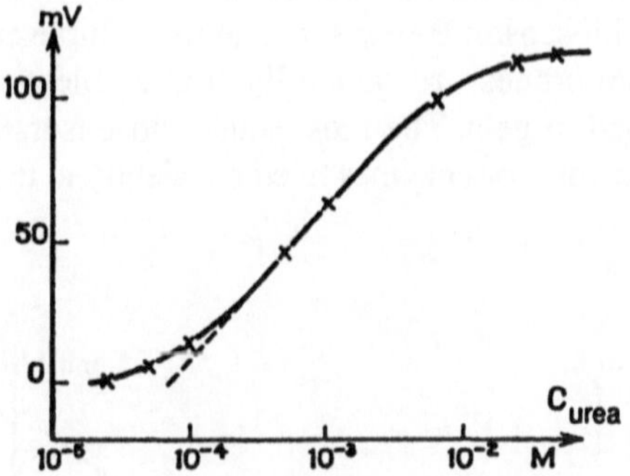

Figure 4.8 *Calibration of a potentiometric enzyme electrode for urea.*

(b) Determination of inhibitors

When an enzyme sensor is intended to determine a species that inhibits an enzymatic reaction, the substrate concentration is fixed at a given value, and the inhibitor concentration is varied [56]. An example of the resulting calibration curve is given in Figure 4.9. In this case, it is possible to represent both the variation in potential and the variation corresponding to the percentage inhibition (I %):

$$I\ \% = \frac{E_0 - E_I}{E_0} \times 100$$

where E_I and E_0 are the potential differences of the biosensor in the presence and absence of inhibitors, for the same concentration of substrate. The advantage of using I % is its dimensionless character.

4.3.3 Parameters affecting enzyme sensor response

The stability of an enzyme sensor depends on a number of physical and chemical parameters. Experimentally, the stability increases with the membrane thickness and enzyme concentration, which is in agreement with the theory set out above. Additional precautions must be taken to ensure the mechanical stability of the enzymatic layer, so that it is always attached to the surface of the transducer and does not tear under agitation. The method used to immobilize the enzyme also affects the stability of the biosensor. Biosensors that use soluble enzymes trapped in dialysis membranes are generally less stable than those using enzymes trapped in gels. The most stable biosensors are those using cross-linked enzymes or enzymes fixed covalently to the transducer.

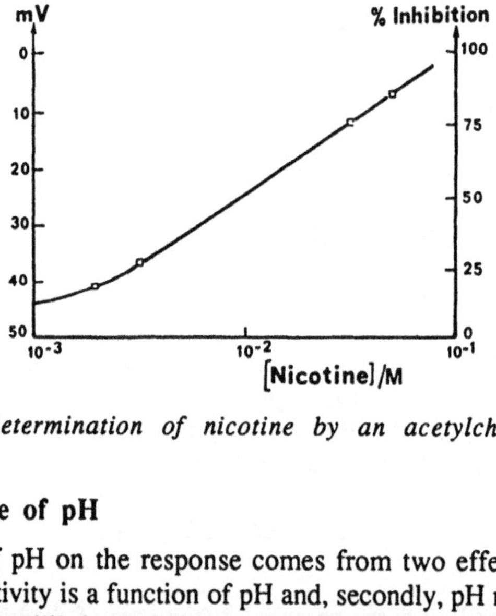

Figure 4.9 *Determination of nicotine by an acetylcholinesterase electrode.*

(a) Influence of pH

The effect of pH on the response comes from two effects. Firstly, enzymatic activity is a function of pH and, secondly, pH may affect a dissociation equilibrium of the product, and only one form will be detectable by the transducer (Figure 4.10). The loss of activity due to the pH can be compensated by increasing the enzyme concentration in the membrane. The conversion of substrate can then be maintained at its maximal level and the enzyme sensor will have very little dependence on pH. The pH-dependence curve is then broader and flatter than that of the same enzyme in solution.

The effect of pH is also observed when the pCO_2 and pNH_3 electrodes are used to determine urea [30]. The pH determines the equilibrium of the dissociation of dissolved CO_2 or NH_3 into bicarbonate or ammonium ions, respectively, and the gas-sensitive

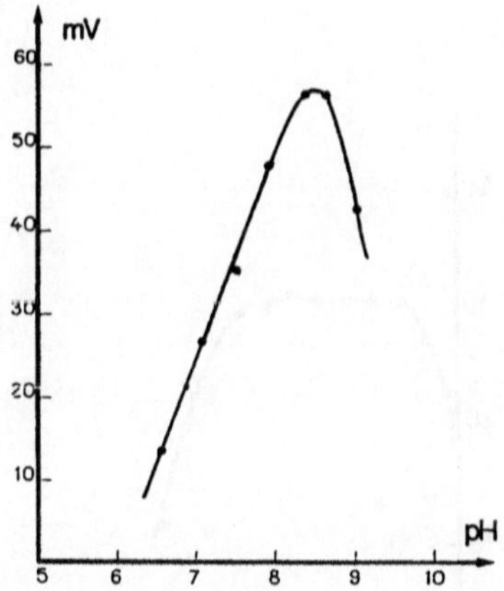

Figure 4.10 *Response of the AChE electrode as a function of pH at 25°C in the presence of 10^{-2} M acetylcholine.*

transducer detects only a fraction of the product obtained in the form of dissolved gas.

(b) Influence of temperature

Sensors use enzymatic reactions and so an increase in temperature will also increase the catalytic activity and hence the rate of reaction. This may, however, be accompanied by denaturation. Temperature also increases the rate of diffusion of the various chemical species in the active layer and improves the response time.

We saw that when the enzyme concentration in the active membrane is high, the concentration of the product formed at the transducer contact is practically the same as that of the substrate in the solution under study. In such a system, the increase in enzymatic activity caused by raising the temperature has little influence on the sensor response.

Figure 4.11 shows that the sensor response is practically unaffected by variations in temperature provided that the temperature range remains between 10 and 40 °C [101]. Above 40 °C, the reduction in enzymatic activity corresponds to the denaturation of the enzyme. Below 10 °C, the apparent activity of the enzymatic layer diminishes, causing a drop in the sensor response. It is advisable to work at around

30 °C, which is 10 °C below the maximal working temperature, but which limits the denaturation thereby prolonging the lifetime of the sensor.

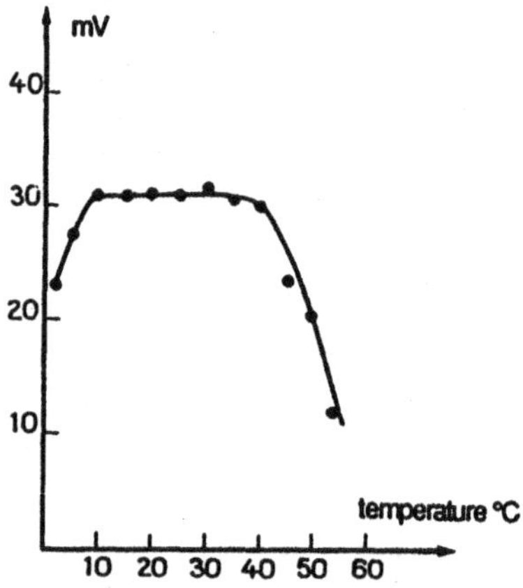

Figure 4.11 *Response of the AChE electrode as a function of temperature in the presence of 10^{-2} M acetylcholine.*

(c) Interference in sensor response

A compound is said to interfere when its presence modifies the response of a sensor in the presence of an analyte. There are three ways a compound may interfere: at the transducer; with the activity of the enzyme; or with the species present during the enzymatic reaction.

Interference at the transducer — Whatever the nature of the transducer, it can never be completely specific to one product. The transducer is much more likely to be selective, which means that it is also sensitive to other substances. The urease electrode uses an electrode that is sensitive to monovalent cations to detect NH_4^+ [99]. This electrode is also sensitive to Li^+, Na^+, K^+ and H^+, hence the idea of using a pCO_2 gas diffusion electrode [30] to eliminate interference from ionic compounds.

Interference with enzymatic activity — Certain compounds increase the activity of the enzyme and are called activators; compounds that reduce this activity are inhibitors. When a substrate is determined in the presence of inhibitors, it is possible to remove the inhibition by increasing the enzyme concentration in the membrane.

Figure 4.12 shows that urea can be determined in blood in the presence of the fluoride ions used to prevent glycolysis after sample collection [102]. An enzyme sensor that uses five units of urease per mm^3 is not affected by fluoride ion concentrations up to 5×10^{-2} M, whereas a concentration of just 3×10^{-3} M will affect the response of the same sensor using only 0.5 units of immobilized urease per mm^3 [56].

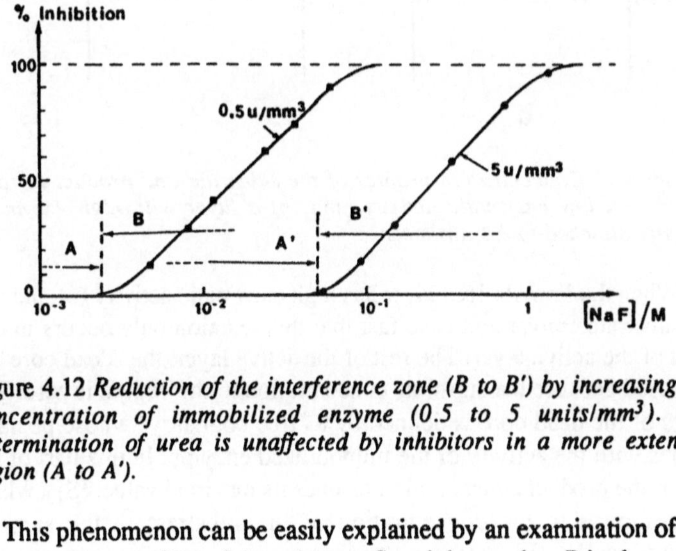

Figure 4.12 *Reduction of the interference zone (B to B') by increasing the concentration of immobilized enzyme (0.5 to 5 units/mm³). The determination of urea is unaffected by inhibitors in a more extended region (A to A').*

This phenomenon can be easily explained by an examination of the concentration profiles of the substrate S and the product P in the active layer on the transducer (Figure 4.13). When the immobilized layer has a low enzymatic activity (a), the substrate concentration diminishes as the substrate diffuses towards the transducer; this concentration equals $[S]_e$ at the transducer. The enzymatic reaction then occurs throughout the entire active layer, and the concentration of the product at the transducer surface is always less than the substrate concentration in the sample ($[P]_e < [S]_0$). In the steady state, where $[S] + [P] = [S]_0$ at all points in the active layer, all the enzyme molecules participate in the reaction. The reaction rate is then given by the value of the product concentration at the surface of the transducer $[P]_e$.

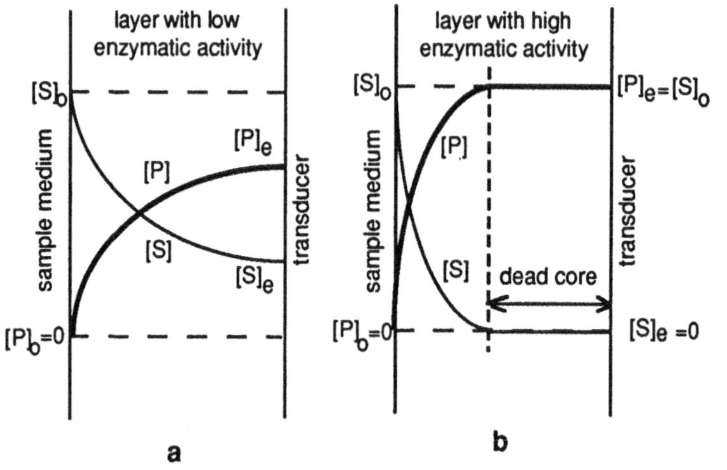

Figure 4.13 *Concentration profiles of the substrate and product in (a) a layer with low enzymatic activity, and (b) a layer with high enzymatic activity attached to a transducer.*

When the immobilized layer has high enzymatic activity (b), the rate of substrate conversion is so fast that this reaction only occurs in one part of the active layer. The rest of the active layer, the "dead core", is not used because the substrate concentration in this region is zero. The edge of the dead core is defined by a "free boundary" whose position varies with the activity of the immobilized enzyme. In this part of the layer, the product concentration reaches its maximal value, $[S]_0$, which corresponds to the concentration of the substrate in the sample. Consequently, $[P]_e = [S]_0 =$ constant, and no variation in enzymatic activity is observable in the presence of activators or inhibitors. Interference by activators and inhibitors can therefore be removed by increasing the enzyme concentration in the active layer.

Interference with enzymatic activity can also arise from the non-specificity of enzymes. The amino acid oxidase sensor is sensitive to most amino acids [3] and can be used for the overall determination of such derivatives. A decarboxylase sensor is much more useful for specific determination of a particular amino acid [103]. The non-specificity of an enzymatic preparation may also be due to contamination by other enzymatic proteins, which can generate a sensitivity towards other species. Purification of the enzyme can eliminate this type of interference.

Interference with the reactants — An error in the measurement may arise if a compound reacts with a cosubstrate or a reaction product. For example, ascorbic acid reacts with oxygen and interferes with the determination of glucose by an amperometric electrode that measures the reduction in the partial pressure of oxygen. This interference is lower when a peroxide-sensitive transducer is used, since this detects the production of H_2O_2.

4.4 Potentiometric enzyme electrodes

An enzyme electrode is constructed by coupling an electrochemical sensor, an electrode, with a thin layer of immobilized enzyme. The enzyme generates a reaction which enables determination of the substrate or the reaction effectors. The electrochemical transducer monitors the enzymatic reaction by potentiometric or amperometric measurements of the product concentration.

Potentiometry is an electrochemical method based on the measurement of the difference in potential between a working electrode and a reference electrode. The determination of the electrode potential gives a direct indication of the concentration of the analyte. The signal obtained is proportional to the logarithm of the concentration (the Nernst Law). The various potentiometric transducers available include the glass electrode for the measurement of pH or monovalent ions, specific electrodes sensitive to anions or cations, and gas electrodes such as the pCO_2 or pNH_3 electrodes. These electrodes are all good candidates for the construction of biosensors; the bioreceptor corresponding to the analyte is simply fixed to the transducer. In an enzyme sensor, one or more enzymes are immobilized on the transducer following the techniques described above (see § 3.3.1).

The potential is measured using a reference electrode with a constant potential that is taken as the origin. The standard calomel electrode (SCE) is generally used for this purpose. The reference electrode is immersed in the solution, next to the enzyme sensor. When a gas electrode is used, the reference component, for example, Ag/AgCl, is placed in the electrolyte on the inside of the enzyme electrode. The enzyme sensor and the reference electrode are connected to a millivoltmeter with a high input impedance ($Z > 10^{12}\ \Omega$), and to electrical circuits which amplify the signal.

4.4.1 Potentiometric urea electrodes

Urea electrodes were among the first potentiometric enzyme sensors. They determine urea in blood and urine, and are based on the enzymatic decomposition of urea by urease:

$$O = C\,(NH_2)_2 + H_2O \xrightarrow{\text{urease}} CO_2 + 2\,NH_3$$

In the pH region where the enzyme is active (around pH 7), the products of the enzymatic reaction above also dissociate:

$$CO_2 + H_2O \rightleftharpoons HCO_3^- + H^+$$
$$NH_3 + H_2O \rightleftharpoons NH_4^+ + OH^-$$

Thus, it is possible to determine urea potentiometrically using a variety of transducers such as the pCO_2 electrode, the pNH_3 electrode, the pH electrode, and the pNH_4 electrode.

(a) pNH_4 transducer

The first transducer used in the construction of a urea sensor was a pNH_4 glass electrode (Figure 4.8) [98, 99]. Urease is physically trapped in a polyacrylamide gel and placed on the tip of the electrode, which is sensitive to ammonium ions. This basic transducer has the drawback of also being sensitive to H^+, K^+, and Na^+ ions, which considerably restricts the use of the sensor in environments containing monovalent cations. The buffer solution is therefore chosen such that its constituents do not interfere with the response of the enzyme electrode [99]. This excludes, for example, the use of a phosphate buffer solution in the calibration of the electrode. Tris-HCl is generally used as the buffer solution which overcomes this problem. Moreover, a calomel reference electrode cannot be used because it contains saturated KCl. The K^+ ions would diffuse across the plug-tube juncture contaminating the sample and perturbing the response of the electrode. This interference can be partly avoided using a Tris-HCl filler solution. Finally, the weak selectivity of the pNH_4 transducer means that this electrode is useless for biological samples such as blood or urine. Interference from Na^+ and K^+ ions at concentrations below 5×10^{-3} and 10^{-4} M, respectively, can be limited by adding a quantity of ion exchange resin (Dowex 50W-X2) to the sample [104]. However, these ions are often present in biological environments at higher (and variable) concentrations, which limits the use of the pNH_4 electrode as a transducer in a urea sensor. The selectivity can be improved somewhat by using another type of pNH_4 electrode in which the antibiotic, nonactin, is incorporated into a PVC membrane [105-106], which forms the sensitive component of the transducer. This modified urea electrode is still too sensitive to K^+ ions.

(b) pH transducer

The pH electrode is another suitable transducer for the construction of a urea biosensor. The classical glass electrode is sensitive to H^+ ions and urease is attached in a gel of either polyacrylamide [107] or methacrylamide-acrylamide copolymer [108]. Some metallic electrodes are also sensitive to H^+ ions (for example, the antimony electrode) and can also be used in conjunction with a urease membrane [109]. The enzyme pH electrode detects a very weak variation in proton concentration arising from an enzymatic reaction, and the signal amplitude is determined by the buffering capacity of the solution. Both the nature of the buffer solution and the working pH value of the biosensor can reduce its practical use. A differential measurement at different pH values can be used to correct for any variation.

The pCO_2 and pNH_3 electrodes thus have the most potential as basic transducers that are specific enough for the construction of a urea electrode [23, 30]. Urease is immobilized on the external face of the hydrophobic membrane to eliminate all passage of ionic compounds. This membrane also increases the response time of the biosensor due to a higher barrier to diffusion.

(c) pCO_2 transducer

The pCO_2 electrode exploits the acidic nature of carbon dioxide, and the dissociation equilibrium that liberates H^+ ions, which are detectable by a glass electrode. It is therefore important to have a solution containing $NaHCO_3$ and NaCl inside the electrode. The $NaHCO_3$ ensures a linear electrode response, and the NaCl stabilizes the Ag/AgCl reference electrode (Figure 4.14).

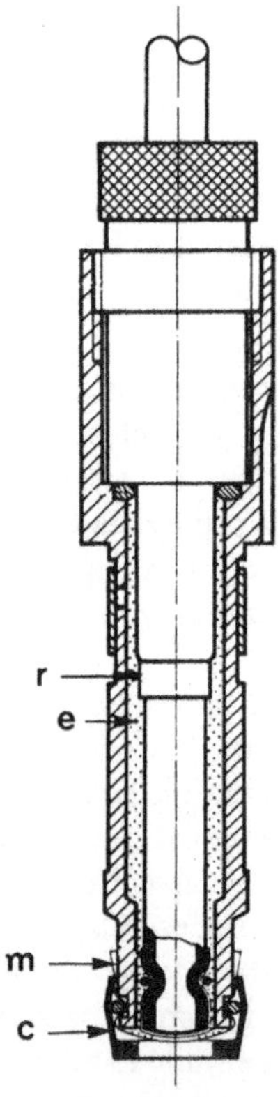

Figure 4.14 *Longitudinal cross-section of a pCO_2 Radiometer electrode which is used as a transducer in a number of biosensors (r - reference electrode; e - intermediate electrolyte solution; m - hydrophobic membrane; c - protective cap).*

When the electrode is covered with a layer of urease (Figure 3.4), the carbon dioxide arising from the decomposition of urea diffuses through the hydrophobic membrane (silicone or teflon) and dissolves in the electrolyte solution. The resulting variation in pH is detectable by a glass electrode. It is thus easy to understand why any other volatile acidic or basic species that can cross the hydrophobic membrane could perturb the electrode response. Fortunately, this type of interference is rare. The presence of CO_2 may cause problems, however, especially if it is present in high concentrations in the sample. The calibration curve of an enzyme electrode that uses the pCO_2 electrode as a transducer (Figure 4.15) has a linear zone which is close to that obtained with the pNH_4 transducer (3×10^{-4} to 10^{-2} M).

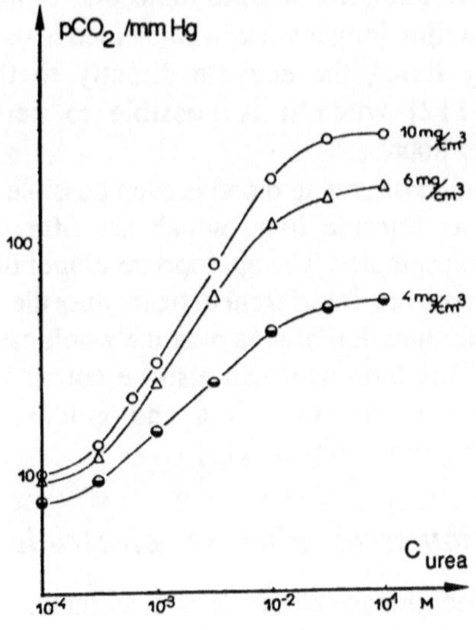

Figure 4.15 *Urea electrode based on a pCO_2 transducer. Influence of the immobilized urease concentration.*

(d) pNH_3 transducer

The pNH_3 electrode exploits the basic properties of ammonia. It has an analogous configuration to the pCO_2 electrode, except for the internal electrolyte of NH_4Cl instead of $NaHCO_3$. The presence of an excess of ammonium ions ensures the linearity of the pNH_3 electrode response. An enzyme urea electrode is constructed using glutaraldehyde to cross-link urease directly onto the surface of a hydrophobic membrane [110]. This reduces the biosensor response time, which varies between 1.5 and 2 minutes for urea concentrations of between 10^{-3} and 10^{-2} M at pH 9, and reaches 4 minutes for 10^{-4} M urea. It is also interesting to note that, although the pNH_3 electrode responds best to ammonium solutions at pH 10, it can still be used as a basic transducer in a urease electrode whose optimal pH is around 7. It is actually the enzymatic activity which plays the determining role, since the best biosensor response is obtained at pH 7. A urea biosensor that uses a pNH_3 or pCO_2 electrode as a transducer can determine urea directly in whole blood, owing to the presence of a hydrophobic membrane which is only permeable to gases. This eliminates the need for pre-treatment of samples to remove interference from molecular or ionic species. The response time is a little longer when whole blood is used [111], but can be improved by fixing the enzyme directly to the hydrophobic membrane [3, 112] when it is possible to perform about 20 measurements per hour.

The determination of urea in blood is even possible when it contains inhibitors such as fluoride ions, which are often used to inhibit glycolysis in blood samples. The appropriate immobilization of urease on the sensor removes interference from fluoride ions [102] and enables direct determination of urea over the whole range of biological concentrations. This biosensor can also be coupled with a glucose oxidase electrode to measure urea and glucose concentrations simultaneously in serum samples [113].

4.4.2 Potentiometric glucose electrodes

The potentiometric glucose electrode is constructed by immobilizing glucose oxidase (GOD) on the sensitive component of a glass pH electrode [114]. The enzyme catalyses the oxidation of glucose into gluconic acid or gluconolactone (the internal ester of gluconic acid):

$$\text{glucose} + O_2 + H_2O \xrightarrow{\text{GOD}} \text{gluconic acid} + H_2O_2$$

The pH electrode detects the gluconic acid formed and is calibrated as a function of glucose concentration. A second enzyme, catalase, is often added to restore some of the oxygen by decomposing the hydrogen peroxide produced. The solution can also be saturated with oxygen to give a maximal response [107].

Glucose oxidase and catalase can also be coimmobilized onto a platinum-iridium wire whose potential is then compared with an Ag/AgCl reference electrode [115]. The electrode response obeys the Nernst law, with a slope of 30.7 mV per decade in glucose concentration. The origin of the potentiometric signal, however, and the mechanism of the electron transfer, have never been elucidated.

4.4.3 Potentiometric amino acid electrodes

The various biosensors constructed for urea used just one enzyme, urease. In contrast, an amino acid biosensor can use a large variety of enzymes, which catalyse the oxidation, hydrolysis, or decarboxylation of a single amino acid.

(a) Oxidases

Amino acids are oxidized by either L-amino acid oxidase (L-AAO) or D-amino acid oxidase (D-AAO), according to their spatial configuration. For example, in the presence of L-AAO, an amino acid from the L-series undergoes the two following reactions:

$$RCH(NH_3^+)COO^- + H_2O + O_2 \xrightarrow{\text{L-AAO}} RCOCOO^- + NH_4^+ + H_2O_2$$

$$RCOCOO^- + H_2O_2 \longrightarrow RCOO^- + CO_2 + H_2O$$

Catalase is often present in oxidases and degrades H_2O_2 as it is formed so that the second reaction cannot occur.

$$H_2O_2 \xrightarrow{\text{catalase}} H_2O + 1/2\ O_2$$

The overall reaction is thus:

$$RCH(NH_3^+)COO^- + 1/2\ O_2 \longrightarrow RCOCOO^- + NH_4^+$$

Amino acid oxidase can thus be used to measure amino acid concentrations amperometrically, using an oxygen electrode, or potentiometrically, using a pNH4 electrode (Figure 4.16). For potentiometric measurements, a glass electrode that is sensitive to monovalent cations is more widely used.

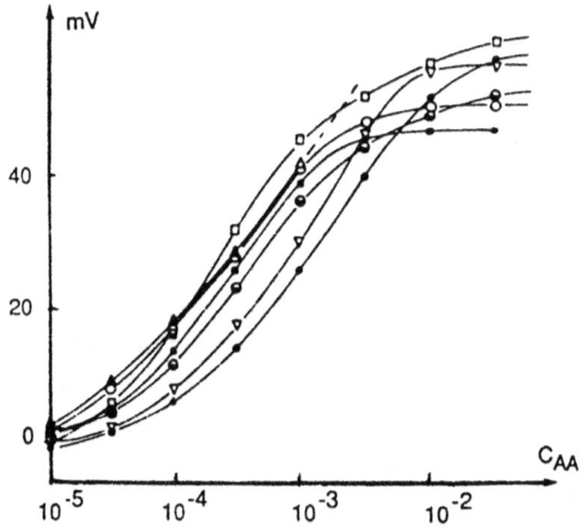

Figure 4.16 *Response of an L-AAO electrode based on the pNH_4 electrode towards different amino acids:* □ *tryptophan,* △ *tyrosine,* ○ *methionine,* ◒ *leucine,* ▽ *arginine,* ● *histidine and* ■ *phenylalanine.*

The enzyme is immobilized by physical entrapment in a polyacrylamide gel [116, 117], or by coreticulation with glutaraldehyde [3]. Physical entrapment requires a nylon net and dialysis paper to ensure the entrapment of the enzymatic layer. The response of the amino acid oxidase electrode towards various amino acids is given in Figure 4.16.

The linear region of the enzyme electrode response varies slightly from one amino acid to the next, but is generally situated between 10^{-5} and 10^{-2} M. The response time is less than one minute when the enzyme is coreticulated directly to the glass electrode [3], and several minutes when the enzyme is attached via other techniques [116, 117]. K^+, Na^+, and Li^+ ions interfere with the electrode response (just like in the urea electrode with the pNH_4 transducer), but this can be avoided with certain precautions. The buffer solution used to calibrate the electrode should not contain these ions, and the reference electrode should not use saturated KCl solution.

(b) Hydrolases

The hydrolysis of an amino acid is catalysed by the relevant amidase which cleaves the amide bond, liberating ammonia. For example, glutaminase hydrolyses glutamine:

$$\text{glutamine} \xrightarrow{\text{glutaminase}} \text{glutamate} + NH_3$$

A glutamine electrode is constructed using a pNH_3 or pNH_4 transducer which exploits the equilibrium between ammonia and ammonium ions. A pNH_4 electrode covered with glutaminase [118] operates at pH 5.5, but the instability of the enzyme limits its useful life to 12 hours. The same enzyme fixed to a pNH_3 electrode gives similar results [119] but a longer response time (4-5 minutes) because the gas has to diffuse through the membrane. The short lifetimes of these glutaminase electrodes makes them very uncompetitive compared with other types of biosensors.

(c) Lyases

Lyases are enzymes that remove certain groups from substrates by mechanisms other than hydrolysis. Decarboxylases and ammonia lyases can be immobilized onto pCO_2 or pNH_3 electrodes, respectively, for amino acid determination. Malate is determined by fixing the malolactic enzyme from *Leuconostoc œnos* to a pCO_2 electrode. The enzyme catalyses the following reaction:

$$\text{malate} \xrightarrow[\text{enzyme}]{\text{malolactic}} \text{lactate} + CO_2$$

This electrode has applications in oenology in the control of malolactic fermentation [120]. Tyrosine is also determined using a pCO_2 electrode, this time covered with tyrosine decarboxylase which catalyses the following reaction [23]:

$$\text{tyrosine} \xrightarrow[\text{decarboxylase}]{\text{tyrosine}} \text{tyramine} + CO_2$$

This enzyme electrode is more specific than those based on deaminases or oxidases, although an excess of decarboxylase is necessary to stabilize the biosensor. Similarly, an electrode based on L-lysine decarboxylase can be constructed to determine lysine in grain or foodstuffs [121], but the response time of this biosensor is long (5-10 minutes). A method using kinetic measurement, based on monitoring the slope of the response curve, has recently been used to reduce the time taken to determine L-lysine [122].

L-phenylalanine is determined using L-phenylalanine ammonia lyase to catalyse the following reaction:

$$C_6H_5CH_2CH(NH_3^+)COO^- \longrightarrow NH_4^+ + C_6H_5CH{=}CHCOO^-$$

$$NH_4^+ \rightleftharpoons NH_3 + H^+$$

A pNH_3 electrode covered with this enzyme is used [123] for the determination of L-phenylalanine with very little interference from other amino acids, notably L-tyrosine. A similar selectivity is obtained with phenylalanine decarboxylase [103] which, when attached to a pCO_2 electrode, enables the specific determination of phenylalanine. Other ammonia lyases are used to determine amino acids. Methionine lyase [124] and histidine ammonia lyase [125] produce ammonia which is detected with a pNH_3 electrode.

Peptides can also be determined potentiometrically using enzyme electrodes. Aspartame, a dipeptide sugar substitute, is determined using a bienzyme electrode, the two enzymes catalysing the following reactions successively [126]:

$$\text{aspartame} \xrightarrow{\text{carboxypeptidase A}} \alpha\text{-cetoglutarate} + \text{L-aspartic acid}$$

$$\text{L-aspartic acid} \xrightarrow{\text{L-aspartase}} \text{fumarate} + NH_4^+$$

The ammonium ions formed are detected with a pNH_3 electrode whose response can be directly related to the aspartame concentration.

4.4.4 Other hydrolase electrodes

Hydrolase electrodes are important because they do not require cofactors. For example, β-galactosidase catalyses the hydrolysis of amygdalin by the following reaction:

amygdalin ---> $C_6H_5CHO + 2\ C_6H_{12}O_6 + HCN$

An amygdalin-sensitive enzyme electrode is obtained by simply immobilizing the enzyme β-galactosidase on a pCN electrode [127]. Similarly, penicillin is determined by cleaving the amide bond of the β-lactame ring with penicillinase to produce penicilloic acid:

$$\text{penicillin} + H_2O \xrightarrow{\text{penicillinase}} \text{penicilloic acid}$$

This acid is detectable by a pH electrode covered with a thin film of penicillinase [128] and the biosensor produced can monitor penicillin concentrations in a variety of fermentation broths (Figure 4.17).

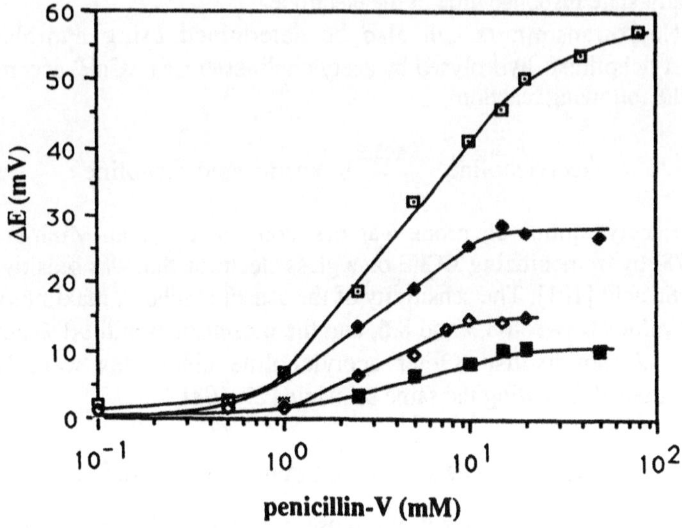

Figure 4.17 *Calibration curves for the penicillinase electrode in (□) 100 % buffer, in (◆) 90 % buffer-diluted, and in (■) undiluted fermentation broth without biomass, and in (◇) fermentation broth with 28 g/l biomass.*

Deaminases hydrolyse certain amines with the liberation of ammonia. One example is adenosine deaminase, which catalyses the following reaction:

$$\text{adenosine} \xrightarrow[\text{deaminase}]{\text{adenosine}} \text{inosine} + NH_3$$

Adenosine is determined by immobilizing the corresponding enzyme on a pNH_3 electrode [129, 130]. It is well known that the pNH_3 electrode, which detects gaseous ammonia (and not the ammonium ion), responds better at basic pH values, and that adenosine deaminase is active over a large pH range. The response of the electrode occurs at much higher pH values (pH 9) than the optimal pH for the enzyme in solution (pH 7). The reaction product inosine is a non-competitive inhibitor of the enzyme and slows the reaction, thereby increasing the steady-state response time of the electrode.

Neurotransmitters can also be determined using hydrolases. Acetylcholine is hydrolysed by acetylcholinesterase (AChE) according to the following reaction:

$$\text{acetylcholine} \xrightarrow{\text{AChE}} \text{acetic acid} + \text{choline}$$

The acetylcholine electrode was first constructed (Tran-Minh *et al.*, 1978) by immobilizing AChE on a glass electrode that was sensitive to acetic acid [101]. The sensitivity of the sensor reaches a maximum for pH values between 8.3 and 8.6, and the biosensor is still 90 % active after 2 months use. Other acetylcholine electrodes were later constructed following the same principle [25, 108].

4.4.5 Inhibitor-sensitive enzyme electrodes

Inhibitors reduce the signal of a biosensor by reducing the activity of the immobilized enzyme on the transducer. This modifies a biosensor calibration curve that is plotted as a function of substrate concentration. As previously described (see § 4.2.4.b), the curve is displaced according to the inhibition mode. The inhibitor is determined by fixing the substrate concentration and varying the inhibitor concentration. A new calibration curve is then plotted as a function of inhibitor concentration. The response of the biosensor indicates whether the inhibition is reversible or not, and inhibitors are classified as reversible or irreversible.

(a) Reversible inhibitors

The reversibility of inhibition is demonstrated by measuring the enzymatic activity before inhibition, then in the presence of the inhibitor, and finally once the inhibitor has been removed. A reference response is obtained by conducting an experiment with the enzyme electrode immersed in a buffer solution containing a predetermined quantity of substrate. The inhibitor is then added in increasing concentrations to determine the percentage inhibition at various concentrations. The reversibility is determined by reducing the inhibitor concentrations, keeping the substrate concentration constant. The same response should be found for the same concentration of inhibitor, and the reference response recovered when all the inhibitor has been removed. Figure 4.18 shows the reversibility of the acetylcholinesterase (AChE) electrode with respect to nicotine (a), and that of the urease electrode with respect to fluoride ions (b).

Some carbamates have therapeutic properties. For example, neostigmine stabilizes acetylcholine at cholinergic synapses using the cholinesterase inhibitor effect. This property can be exploited to determine neostigmine [131] with an AChE biosensor that responds linearly to concentrations between 10^{-6} and 10^{-5} M.

Effect of enzyme concentration — Enzymatic inhibition occurs via a well-defined chemical coupling. The rate of the enzyme-catalysed reaction is then a function of the concentration of the enzyme with respect to that of the inhibitor. As the enzyme concentration increases, the inhibitor concentration must also increase to ensure the same degree of inhibition (Figure 4.12).

At each enzyme concentration, there is a corresponding calibration curve, whose linear region can be used for inhibitor determination. As the immobilized enzyme concentration increases, the linear region of the curve undergoes a translation towards higher inhibitor concentrations (NaF). Consequently, the detection limit of a biosensor designed for inhibitor determination drops with the concentration of immobilized enzyme. In reality, the residual activity of the enzyme after immobilization determines the electrode response [132]. This can be

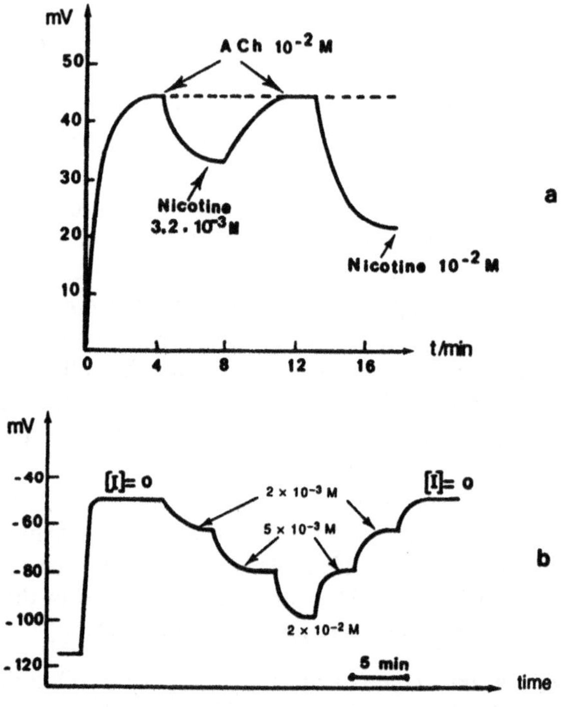

Figure 4.18 *Enzyme electrodes that are sensitive to reversible inhibitors, (a) the AChE electrode for nicotine determination, and (b) the urease electrode for the determination of fluoride ions.*

verified by immobilizing the same quantity of enzyme with different concentrations of cross-linking agent and different binding times. When glutaraldehyde is used as the cross-linking agent at high concentrations and with long binding times, the enzyme is partly denatured, which induces a reduction in the activity of the immobilized enzyme and a displacement of the curves towards weak inhibitor concentrations. Inhibitor measurement thus requires predetermination of the method and conditions of enzymatic immobilization so as to give optimal operation of the biosensor in the inhibitor concentration range desired [133]. The best methods are those using batch preparation of membranes because they provide a better reproducibility in the measurement.

Effect of substrate concentration — The influence of substrate concentration on the response of an enzyme electrode in the presence of an inhibitor varies according to the inhibition mode (see § 4.2.4.a). The degree of inhibition for an uncompetitive inhibitor, like NaF on urease, increases with the substrate concentration for a given biosensor (Figure 4.19). The detection limit for fluoride ions is thus improved by using high urea concentrations [132].

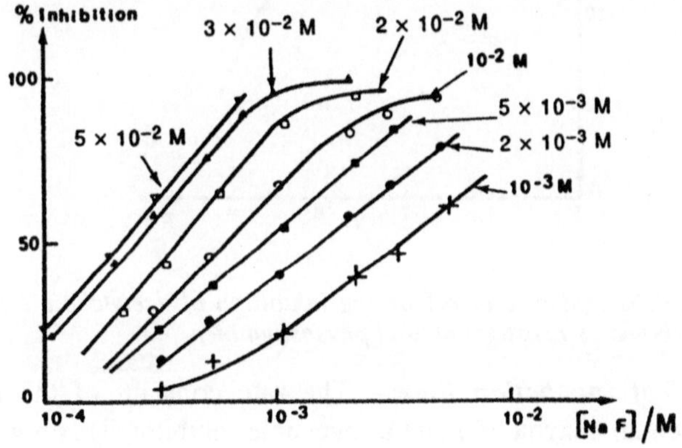

Figure 4.19 *Influence of the substrate (urea) concentration on the calibration curve of an enzyme (urease) electrode designed for the determination of uncompetitive inhibitors (NaF).*

Effect of pH — Enzymes are generally only active in limited pH ranges, and each enzyme has its own optimal pH value. The presence of an inhibitor changes the different ionization states of the system governing the enzymatic activity, which results in a variation in the behavior of the inhibited enzyme with respect to pH from one inhibitor to another. This phenomenon can be studied using an immobilized enzyme electrode, the response of which gives the enzymatic activity in the presence of various inhibitors (Figure 4.20). The inhibition of AChE by neostigmine is favored by high pH values, which is not the case for physostigmine. Furthermore, the kinetic parameters V_m and K_m are pH dependent, and so the influence of pH on enzymatic inhibition also varies according to the inhibition mode.

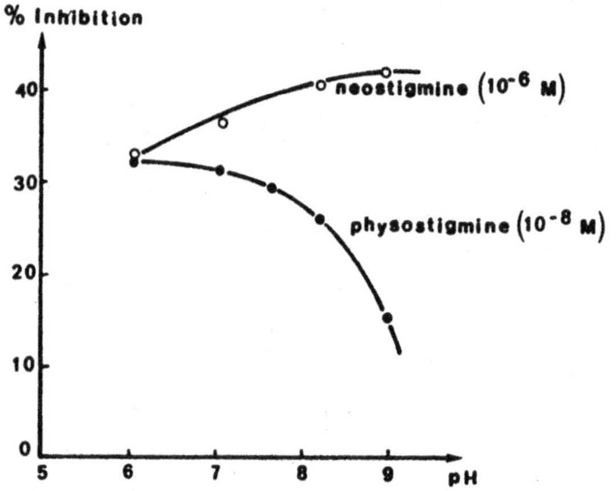

Figure 4.20 *Influence of pH on the inhibition of acetylcholinesterase by the carbamates neostigmine and physostigmine.*

Effect of incubation time — The rate constant of the reaction between an enzyme (E) and a reversible inhibitor (I) is high and an equilibrium state with the complex EI is rapidly achieved. Therefore, the steady-state biosensor response for a given reversible inhibitor concentration does not depend on the incubation time [132] when the enzyme is in contact with the inhibitor before the substrate is added (Figure 4.21). This does not apply for irreversible inhibitors, which require a much longer time to reach an equilibrium state with the corresponding enzyme.

(b) Irreversible inhibitors

Irreversible inhibitors can have a fatal effect on an organism through their action on the enzymes in the metabolic chain. The ever-growing discharge of toxic products into nature leads to the requirement for methods that can monitor them in low concentrations for their control in the environment. Heavy metals and pesticides are important examples of such materials.

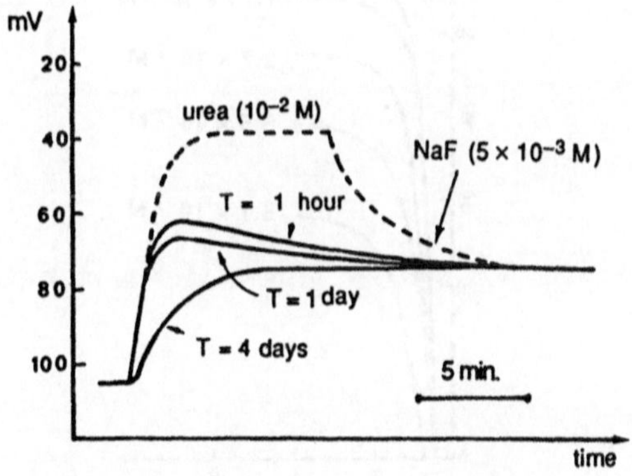

Figure 4.21 *Effect of incubation time, T, on the response of a urease electrode that is sensitive to fluoride ions. The dotted line represents the response of the enzyme sensor with respect to its substrate without incubation, before and after addition of fluoride ions.*

Heavy metals — High concentrations of heavy metals lead to the precipitation of proteins and the denaturation of enzymes. At lower concentrations, they affect enzymatic activity through their interaction with the thiol and carbonyl groups of the enzyme. Biosensors exploit this property for the determination of heavy metals [56, 95].

Mercury ion levels are determined by immobilizing urease on a pCO_2 electrode. This must be done in the absence of albumin to avoid the reaction between the albumin thiol groups and mercury. The response curves (Figure 4.22) are obtained after 10 minutes incubation when the response to the substrate (10^{-2} M urea) is tested.

The incubation technique allows inhibition in spite of the low rate of complexation between the enzyme and its inhibitor. The percentage inhibition indicated by the biosensor signal is a function of the incubation time. An increase in incubation time leads to a higher degree of inhibition and favors the detection of the inhibitor at low concentrations.

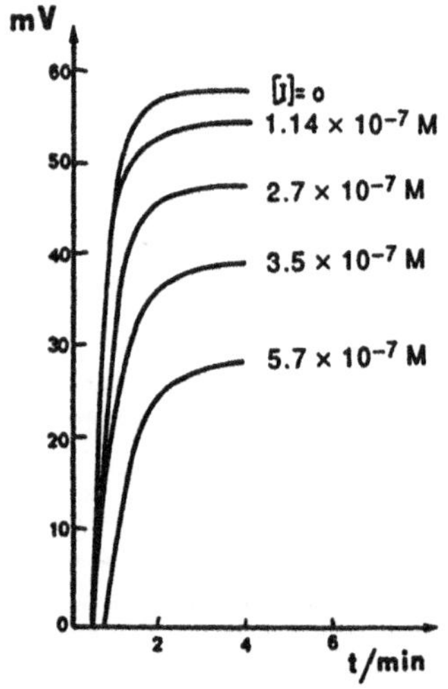

Figure 4.22 *Response curves of an enzyme sensor that is sensitive to different concentrations of mercury ions.*

When the inhibitor is a heavy metal, complexation occurs at the thiol groups, according to reactions of the type:

$$E\ SH + M^{+} \rightleftharpoons E\text{-}SM + H^{+}$$

$$E\ (SH)_2 + M^{2+} \rightleftharpoons E\ (S)_2M + 2H^{+}$$

It follows that the fixation of a heavy metal to an enzyme is a function of pH. The inhibition increases with pH, whereas the reactivation of the enzyme is favored by decreasing values of pH [56]. The enzyme can be reactivated very quickly by precipitation or by complexation of the heavy metal with agents such as KI, $(NH_4)_2S$ or EDTA.

When a number of metals are present in the sample medium, the biosensor gives a response that corresponds to overall inhibition by the various metals. The biosensor is reactivated in several steps by selective complexing agents. This operation gives a measure of the individual concentrations of the different metals using their corresponding calibration curves (Figure 4.23).

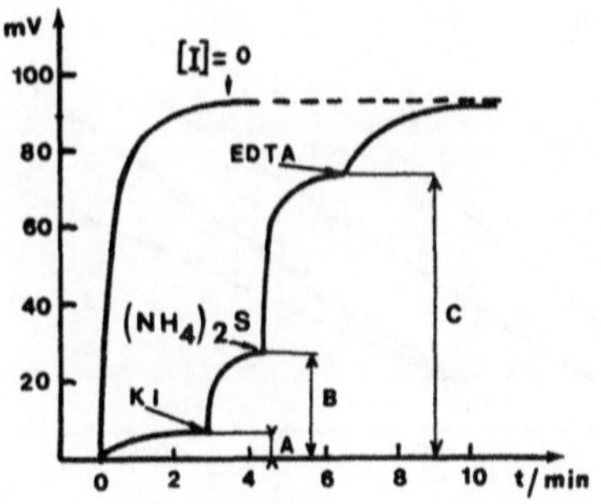

Figure 4.23 *Determination of several metals present in the same solution ($[Co^{2+}] = 10^{-6}$ M, $[Ag^{+}] = 10^{-6}$ M, $[Hg^{2+}] = 5\times 10^{-7}$ M) using a urease electrode which is inhibited and reactivated in several steps by the selective complexing agents KI, $(NH_4)_2S$ and EDTA.*

Pesticides — Pesticides are synthesized to control animal and plant parasites, for example, the organophosphates that are used as agricultural insecticides. Chemical warfare agents come under this classification because they also provoke death by respiratory paralysis. The action of these compounds is based on the inhibition of the enzymes in the nervous system. Organophosphate derivatives react with acetylcholinesterase (AChE) to form very stable phosphoryl-enzyme complexes. This reaction may occur at very low organophosphate levels, and so a method that measures concentrations of the order of ppb (parts per billion) is necessary. This can be achieved by immobilizing the target enzyme on electrochemical transducers. A glass electrode with immobilized AChE is sensitive to the various pesticides that inhibit this enzyme (Figure 4.24).

Inhibition by organophosphates is a slow process and thus requires incubation to improve the detection limit. A very low pesticide concentration requires a longer incubation time to produce the same degree of inhibition. For the same inhibitor concentration, the loss in enzymatic activity follows first order kinetics (Figure 4.25).

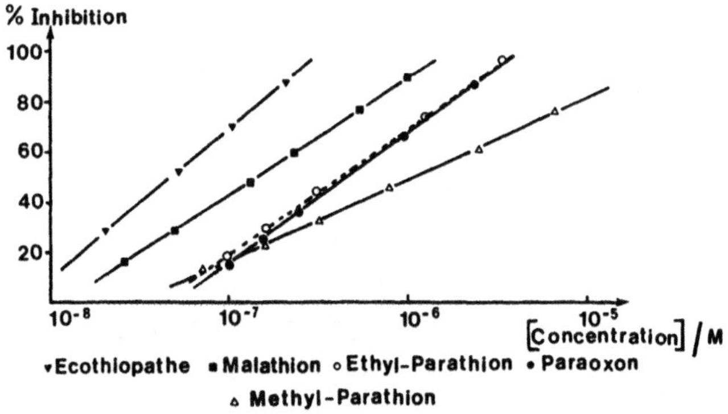

Figure 4.24 *Calibration curves for the acetylcholinesterase electrode with various pesticides.*

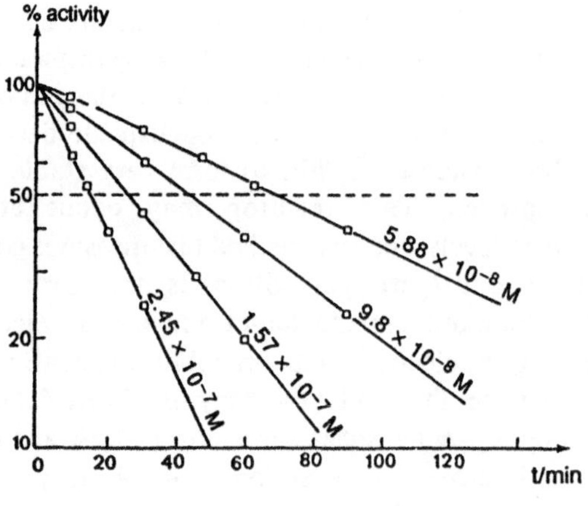

Figure 4.25 *Enzyme sensor monitoring of the inhibition kinetics of AChE in the presence of different concentrations of paraoxon.*

The spontaneous reactivation of the inhibited enzyme is very slow. Reactivators, such as oximes, must be used for rapid restoration of the original activity of the enzyme. Both inhibition and reactivation occur at the serine group (Ser) at the active site of the enzyme (Enz). These processes are described by the following reaction scheme:

$$\text{Enz-Ser-OH} + \text{phosphate} \xrightarrow{\text{inhibition}} \text{Enz-Ser-phosphate}$$

$$\text{PAM} + \text{Enz-Ser-phosphate} \xrightarrow{\text{reactivation}} \text{Enz-Ser-OH} + \text{PAM–phosphate}$$

The reactivation is more effective with pyridine aldoxime-methyl (PAM) than with hydroxylamine; 10^{-4} M PAM reactivates at the same rate as 10^{-2} M hydroxylamine. Figure 4.26 shows the biosensor response in the absence of pesticides (a), and then in the presence of pesticides, before and after addition of PAM (b).

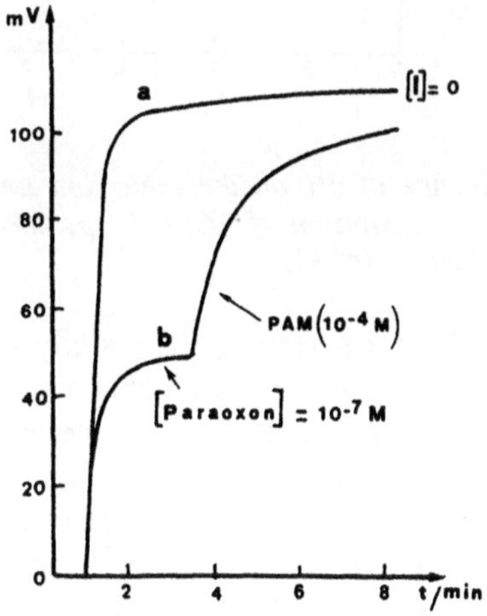

Figure 4.26 *Regeneration of an AChE electrode by PAM, (a) response in the absence of inhibitor, and (b) response in the presence of paraoxon before and after addition of PAM.*

The inhibition and regeneration of the cholinesterase electrode exhibit different variations with changing pH (Figure 4.27), as was the case for heavy metals. The inhibitor action of paraoxon is optimal for a pH of about 7.4 and so the serum environment favors organophosphate intoxication. Regeneration of the enzyme is favored for more basic pH values [134].

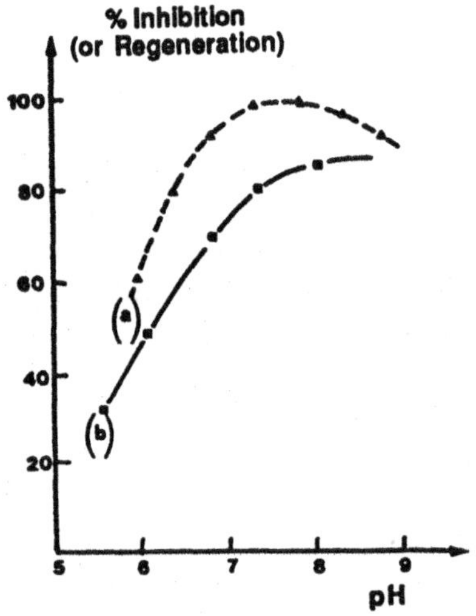

Figure 4.27 *Influence of pH on the inhibition and regeneration of an enzyme sensor, (a) inhibition of AChE by paraoxon (10^{-6} M), and (b) regeneration by PAM (10^{-4} M).*

Measurement of total toxicity — The study of inhibition showed us that a single enzyme may be the target of a number of different toxic compounds with a large variety of characteristics. This observation leads to the idea of constructing a sensor capable of measuring total toxicity [135]. Such a biosensor could replace the biological tests currently used (for fish and daphnia mortality, and bacteria bioluminescence, Microtox). A range of toxic products must be chosen that is representative of all the different structures, such as methylazinphos, paraquat, 2,4,6-trichlorophenol, carbon tetrachloride, and lindane.

Generally, the evaluation of the enzyme-inhibiting power of toxic compounds involves the measurement of the concentration of toxic product required for the inhibition of 50 % of the enzymatic activity. This is called the CI_{50}. When the detection limit of a biosensor is required then the CI_{10} can be used, which corresponds to an inhibition of 10 % of the enzymatic activity.

Table 4.1 gives the detection limits of toxic compounds with the enzyme acetylcholinesterase (AChE) from the electric eel, compared with the results of luminescence assays on "Microtox" carried out by the *Agence de Bassin RMC*, using the same toxic compounds [135]. The AChE biosensor has about the same detection limit for methylazinphos as Microtox. Its detection limit is lower for paraquat, at 3 ppm (parts per million), and much higher for the other toxic compounds.

Method \ Toxic agent	Paraquat	Trichloro-phenol	Methyl azinphos	Carbon tetrachloride	Lindane
AChE electrode (electric eel)	3 ppm	20 ppm	150 ppb	-	15 ppm
Microtox	500 ppm	500 ppb	200 ppb	3 ppm	1 ppm

Table 4.1 *Detection limits for various toxic agents detected by an enzyme electrode, or by the "Microtox" method (the concentrations correspond to an inhibition to 10 % of the enzymatic activity).*

Remark — The reduction in enzymatic activity in the presence of an inhibitor has served as a good base for the construction of inhibitor-sensitive electrodes. It is obvious that any compound whose presence modifies the biosensor signal, can be determined by that biosensor. This is also the case for *activators*, which *increase* the response signal of the transducer. In practice, however, this application is of limited importance.

4.4.6 Comparison of potentiometric enzyme electrodes

A number of potentiometric enzyme electrodes are listed in Table 4.2 to provide an overall view. The components required, the enzyme, and the transducer are listed for each analyte that is detectable by biosensors. The performance of the electrodes are presented using two essential parameters, the response time and the range of concentrations in which the signal obtained is linear as a function of the logarithm of the concentration. More detailed descriptions of each biosensor can be found in the references indicated [23-137]. Immobilized enzyme measuring systems that use microcolumns associated with electrochemical detection are excluded from this table [138] because they are not biosensors but analytical techniques.

The first method for immobilizing enzymes on transducers via coreticulation of the enzyme and a load protein (bovine serum albumin) using glutaraldehyde was proposed in 1972 by Tran-Minh *et al.* [99]. Tables 4.2 and 4.3 show that this method has often been employed in the construction of biosensors, owing to its simplicity and its rapidity.

4.5 Amperometric enzyme electrodes

Amperometry involves the determination of the intensity of a current as it flows through an electrochemical cell at an applied potential. This intensity is a function of the concentration of the electroactive species being oxidized or reduced at a probe electrode, a second electrode acts as a reference. After proper calibration, these intensities can be used to measure the concentration of some of the species present.

When an amperometric electrode is used as the transducer of a biosensor, there is a consumption of reaction products; this is the major difference from a potentiometric electrode. The diffusion-reaction equations ((3) and (4)) still apply, assuming that the product concentration at the transducer-active layer interface is zero ([P] = 0). This hypothesis corresponds to the maximal sensitivity of the biosensor. The flow of product towards the transducer can be limited by mass transfer, either in the semi-permeable membrane, which separates the enzyme from the sample medium [139], or in the active enzymatic layer [140]. When the semi-permeable membrane is the limiting factor of a diffusion process, all of the product formed is

Table 4.2(a) *Some examples of potentiometric enzyme sensors*

Analyte	Enzyme (E)	Immobilization technique	Transducer	Response time	Linearity	Reference
urea	urease	E + polyacrylamide	pNH_4	25 - 60 s	5×10^{-5} - 10^{-2} M	[98]
urea	urease	E + BSA + GA	pNH_4	30 s - 1 min	10^{-4} - 10^{-2} M	[3, 99]
urea	urease	E + polyacrylamide	pNH_4 (nonactine)	30 - 40 s	10^{-5} - 2×10^{-2} M	[105]
urea	urease	E + polyacrylamide	pH (glass)	5 - 10 min	5×10^{-5} - 5×10^{-4} M	[107]
urea	urease	E + copolymer methacrylamide-acrylamide	pH	3 - 5 min	2×10^{-5} - 3×10^{-4} M	[108]
urea	urease	E + BSA + GA	pH (antimony)	1 - 1.5 min	2×10^{-4} - 2×10^{-3} M	[109]
urea	urease	soluble E / dialysis membrane	pCO_2	1 - 5 min	10^{-4} - 10^{-1} M	[23]
urea	urease	E + BSA + GA	pCO_2	2 - 3 min	3×10^{-4} - 10^{-2} M	[30]
urea	urease	E + BSA + GA	pCO_2	1 - 2 min	3×10^{-4} - 10^{-2} M	[3]

Table 4.2(b)

urea	urease	E + GA	p NH_3	1.5 - 2 min	10^{-4} - 10^{-2} M	[110]
urea	urease	soluble E / cellophane membrane	p NH_3	5 - 8 min	5 x 10^{-4} - 7 x 10^{-2} M	[111]
urea	urease	E + BSA + GA	p NH_3	2 - 3 min	3 x 10^{-4} - 10^{-2} M	[112]
urea	urease	E / nylon	p NH_3	26 - 38 s	0.2 - 10 mg / dl	[113]
glucose	glucose oxidase	E + BSA + GA	Pt - Ir	-	1 - 150 mg/ml	[115]
amino acids	L-amino acid oxidase	E + polyacrylamide	p NH_4	3 - 5 min	2 x 10^{-4} - 2 x 10^{-3} M	[116]
amino acids	D-amino acid oxidase	E + polyacrylamide	p NH_4	1 - 2 min	10^{-4} - 5 x 10^{-3} M	[117]
amino acids	L-amino acid oxidase	E+ BSA + GA	p NH_4	30s -1 min	10^{-4} - 10^{-2} M	[3]
glutamine	glutaminase	E / nylon / cellophane	p NH_4	1 - 2 min	10^{-4} - 10^{-1} M	[118]
glutamine	glutaminase	E / dialysis membrane	p NH_3	4 - 5 min	10^{-4} - 2 x 10^{-3} M	[119]
tyrosine	tyrosine decarboxylase	E / dialysis paper	p CO_2	1 - 5 min	3 x 10^{-4} - 5 x 10^{-3} M	[23]

Table 4.2(c)

lysine	lysine decarboxylase	E + BSA + GA	p CO_2	5 - 7 min	10^{-4} - 3×10^{-3} M	[121]
phenylalanine	phenylalanine ammonia lyase	E / microchamber	p NH_3	12 min	10^{-4} - 6×10^{-4} M	[123]
methionine	methionine lyase	E + BSA + GA	p NH_3	3 min	10^{-5} - 10^{-2} M	[124]
histidine	histidine ammonia lyase	E + BSA + GA	p NH_3	3 - 8 min	3×10^{-5} - 10^{-2} M	[125]
aspartame	carboxypeptidase A L-aspartase	E + BSA + GA	p NH_3	4 - 8 min	4.25×10^{-4} - 8.1×10^{-3} M	[126]
amygdalin	ß-glucosidase	E + polyacrylamide	p CN	10 - 30 min	10^{-5} - 10^{-2} M	[127]
penicillin	penicillinase	E + polyacrylamide	pH	2 - 3 min	4×10^{-5} - 10^{-3} M	[108]
adenosine	adenosine deaminase	E + BSA / dialysis membrane	p NH_3	7 - 12 min	7×10^{-7} - 10^{-2} M	[129]
adenosine	adenosine deaminase	E + BSA + GA	p NH_3	5 - 14 min	3.5×10^{-5} - 1.6×10^{-2} M	[130]
acetylcholine	acetyl-cholinesterase	E + BSA + GA	pH	30 s - 1 min	10^{-4} - 10^{-2} M	[101]

Table 4.2(d)

acetylcholine	acetylcholinesterase	E + GA	pH	1 min	10^{-4} - 3×10^{-3} M	[25]
acetylcholine	acetylcholinesterase	E + polyacrylamide	pH	1 - 3 min	2×10^{-5} - 10^{-3} M	[108]
nicotine	acetylcholinesterase	E + BSA + GA	pH	2 - 3 min	10^{-3} - 10^{-1} M	[131]
neostigmine	acetylcholinesterase	E + BSA + GA	pH	2 - 3 min	5×10^{-7} - 5×10^{-5} M	[131]
physostigmine	acetylcholinesterase	E + BSA + GA	pH	2 - 3 min	3×10^{-8} - 3×10^{-6} M	[131]
fluoride	urease	E + BSA + GA	$p\,CO_2$	2 - 5 min	3×10^{-4} - 10^{-2} M	[132]
fluoride	acetylcholinesterase	E + BSA + GA	pH	2 - 3 min	10^{-4} - 2×10^{-2} M	[26]
Ag^{+}	urease	E + BSA + GA	$p\,CO_2$	1 - 2 min	10^{-8} - 3×10^{-7} M	[95]
Hg^{2+}	urease	E + GA	$p\,CO_2$	1 - 2 min	2×10^{-8} - 2×10^{-6} M	[56]
pesticides	cholinesterase	E + BSA + GA	pH	2 - 3 min	10^{-8} - 10^{-5} M (according to pesticide)	[56]
H_2O_2	peroxidase	E + BSA + GA / collagen	p F	–	8×10^{-5} - 10^{-2} M	[136]
arginase	urease	E + polyacrylamide	$p\,NH_4$	< 10 min	1.6 - 16 units	[137]

detected by the transducer. In the steady state, the reaction rate is equal to the rate of diffusion of the substrate in the semi-permeable membrane. A very permeable membrane will therefore enable the measurement of low substrate concentrations, whereas a membrane that is not very permeable will require high concentrations of substrate for detection. When the mass transfer is limited by diffusion in the active enzymatic layer, the situation corresponds to the general case described in the theoretical section above (see § 4.2). The presence of a very permeable membrane does not affect the biosensor at all, its behavior being related essentially to the kinetics of diffusion and reaction in the enzymatic layer. It is therefore possible to obtain the theoretical response of an enzyme sensor in the steady state, or in a transient state [140], even in the presence of a cofactor [141].

4.5.1 Glucose sensitive electrodes

Glucose sensitive electrodes are needed to monitor and regulate the concentration of glucose in the blood of diabetic patients. Glucose is also the most important substrate in the fermentation industry. In 1962, Clark and Lyons had the idea of determining glucose by immobilizing a layer of glucose oxidase on a pH or pO_2 electrode [1]. The pH electrode detects the gluconic acid produced by the reaction, and the O_2 electrode measures the consumption of oxygen by the enzyme. In 1967, Updike and Hicks immobilized glucose oxidase (GOD) in a polyacrylamide gel on the tip of an oxygen-sensitive electrode (pO_2), and plotted a calibration curve for the enzyme electrode as a function of glucose concentration. The overall reaction is:

$$\text{glucose} + O_2 + H_2O \xrightarrow{\text{GOD}} \text{gluconic acid} + H_2O_2$$

The consumption of oxygen is proportional to the rate of the enzymatic reaction, and hence the glucose concentration, especially when this is small compared with the Michaelis constant of the enzyme (see § 4.2). The partial pressure of oxygen can be monitored amperometrically, and so this constitutes a possible method for glucose determination. Glucose can also be determined from the products of the enzymatic reaction: gluconic acid (via the potentiometric methods described in § 4.4.2) and hydrogen peroxide (via amperometric methods).

(a) Measurement of oxygen partial pressure

The partial pressure of oxygen is measured using an electrochemical minicell in which the oxygen is reduced to a constant potential equal to –0.65 V/SCE (with respect to the standard calomel electrode (SCE)). The pO_2 electrode (or Clark electrode) has a platinum cathode and an Ag/AgCl anode, which are immersed in an electrolytic solution. The whole system is covered with an oxygen-permeable hydrophobic membrane. This hydrophobic membrane serves as a barrier to all the electroactive species that could come into contact with the electrodes and disturb the measurement. Oxygen is reduced by the action of the potential, and the oxygen concentration drops inside the electrochemical cell. The oxygen partial pressure on the exterior is constant, and so an oxygen concentration gradient is set up in the membrane, which induces a flow of oxygen towards the platinum electrode. The current received at the electrode is proportional to the oxygen concentration gradient, and hence the oxygen pressure in the sample solution (Fick's Law). In order to prevent the exhaustion of the oxygen in the sample, the size of the platinum cathode is reduced and a wire with a diameter of 10 to 25 μm is used. The reference electrode, Ag/AgCl, is much larger in comparison. The transfer of oxygen from the exterior to the interior of the electrode is slow because it is diffusion controlled. Consequently, the response time of the pO_2 electrode is about one minute, but varies according to the thickness of the membrane used (silicone, PTFE, etc.). An enzyme electrode that uses a pO_2 transducer can never have a shorter response time than this.

When a pO_2 electrode is covered with a layer of glucose oxidase and placed in a sample that contains glucose, oxygen is consumed in two ways. Some of the oxygen is used by the enzymatic reaction that oxidizes the glucose, and some of it diffuses to the interior of the pO_2 transducer, where it is electrochemically reduced. Considering the size of the reducing surface of the cathode, it can be assumed that the enzymatic reaction uses most of the oxygen. The rate of this reaction is a function of the oxygen concentration, and so the response of the biosensor to glucose [3] depends on the initial partial pressure of oxygen in the sample (Figure 4.28). This also explains why the detectable glucose concentration range increases with the partial pressure of oxygen. A stable oxygen concentration will ensure proper operation of the enzyme electrode. This can be achieved either by using samples with a large volume, or by using an external supply of oxygen. The presence of reducers, such as ascorbic acid, must be avoided.

The most convenient way of reducing the exhaustion of oxygen in the sample is to coimmobilize glucose oxidase and catalase [3], which catalyse the following reactions:

$$\text{glucose} + O_2 + H_2O \xrightarrow{\text{GOD}} \text{gluconic acid} + H_2O_2$$

$$H_2O_2 \xrightarrow{\text{catalase}} H_2O + 1/2\, O_2$$

The enzymatic decomposition of H_2O_2 can thus be used to retrieve half of the oxygen consumed, which means that the oxygen is consumed half as quickly.

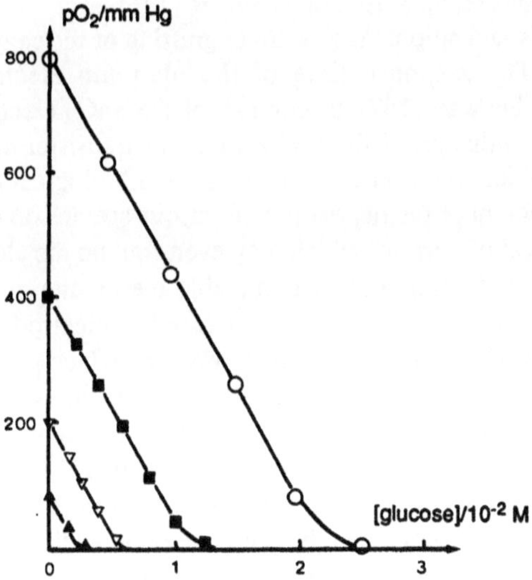

Figure 4.28 *Calibration curves of a glucose-sensitive electrode, based on the pO_2 transducer, as a function of oxygen partial pressure in the sample, between 100 and 800 mm Hg.*

The pO_2 transducer is useful because it provides a hydrophobic support onto which an enzyme can easily be fixed. Glucose oxidase can be immobilized trapped in a polyacrylamide gel [2, 142], coreticulated with albumin [3] or gelatine [143], or immobilized using a nylon net [113]. Enzyme electrodes obtained in this way have been used for the determination of glucose in blood and serum, and have also been tested for monitoring fermentation processes [144].

(b) Measurement of hydrogen peroxide concentration

Hydrogen peroxide can be detected amperometrically using a transducer constructed in a similar way to the pO_2 electrode. The O_2-permeable hydrophobic membrane is replaced by a hydrophilic membrane that is permeable to H_2O_2. The hydrogen peroxide can then reach the platinum electrode. The hydrogen peroxide is oxidized by polarizing the platinum electrode to a positive potential, +0.6 V/SCE, which makes it an anode. This working electrode can be pre-treated by submitting it to a cyclic potential to give a low detection limit [145], or a differential measurement can be made with an enzyme-free electrode to eliminate interference from other sugars [146].

Electrode silanization enables direct grafting of the enzyme onto the transducer. The response time of the platinum electrode can be increased in this way [147], as can that of the SnO_2 electrode, which, being a semiconductor, facilitates its use in the transistor mode [148].

The use of an amperometric electrode in a biological environment poses the problem of the deposition of various species on the electrode under the effect of current, which may even foul up the electrode. This can be prevented using a semi-permeable membrane made of either cellulose acetate, which is fixed to a rotating electrode [149], or a composite material (a multi-layered membrane), which gives a selective permeability [150]. It is also possible to cover the whole assembly (cathode and enzyme anode) with a thin film of a silicone elastomer that is permeable to glucose. The biosensor obtained [151] may be used in whole blood with a very extended dynamic range (up to 6 g/l of glucose). It is well known that the presence of an additional membrane gives rise to a resistance to transfer, and imposes a concentration gradient. The transducer thus works at much lower analyte concentrations than are present in the sample. This technique means that the upper limit of the range of detectable concentrations can be pushed even higher [152]. It must, however, be recalled that the sensitivity is reduced by the same proportion, since this is represented by the ratio dI/dc, the variation in the signal with that of the concentration [100].

An H_2O_2-sensitive transducer can be used for glucose determination in the place of a pO_2 electrode, but this does not eliminate the dependence of the biosensor on oxygen, which is still required for the enzymatic oxidation of glucose. If there is no continuous, and stable, supply of oxygen (via an air supply [153, 154] or electrolysis [155]) to compensate for the consumption of the enzymatic reaction, then the signal emitted by the electrode does not give a correct reading of the glucose concentration. There is an increasing need for sensors that are sensitive to glucose but which operate independently of oxygen concentration. Such biosensors could be implanted in living tissue where the oxygen content varies a great deal, notably in the brain, or installed *in situ* in fermenters, where the oxygen concentration may be zero. Two approaches have been chosen in the construction of such biosensors. One method involves using mediators, and the other involves the direct transfer of the cofactor electrons to the electrode, in order to ensure the regeneration of the enzymes.

(c) Use of mediators

The overall reaction for the oxidation of glucose described above actually uses an enzyme cofactor FAD (flavin adenine dinucleotide). When this prosthetic group is bound to glucose oxidase, it ensures the transfer of electrons from the substrate, and hence the oxidation of that substrate. After oxidation, the cofactor FAD exists in its reduced form ($FADH_2$) and returns to its oxidized form (FAD) in the presence of oxygen as indicated in the following reaction scheme:

$$\text{glucose} + \text{GOD(FAD)} \longrightarrow \text{gluconic acid} + \text{GOD}(\text{FADH}_2)$$

$$\text{GOD}(\text{FADH}_2) + O_2 \longrightarrow \text{GOD(FAD)} + H_2O_2$$

This oxygen dependence can be eliminated by choosing a convenient method of reoxidizing the cofactor, $FADH_2$. Since amperometry uses electrodes and a source of potential, it may be possible to transfer the cofactor electrons directly to the electrode to regenerate the enzyme. In practice, this transfer can only occur when the enzyme is soluble [156] because its mobility then allows the prosthetic group, FAD, to make contact with the electrode. When the enzyme is immobilized on the transducer, the FAD is trapped at the active site of the enzyme, and cannot make contact with the electrode, which is situated at some distance. A mediator is necessary to transfer the electrons by acting as a bridge or a shuttle between the cofactor and the electrode. The mediator is reduced to M_{red} by the cofactor, and then oxidized into M_{ox} when it comes into contact with the anode, which is polarized at the appropriate potential. The mediator thus reacts according to the following scheme:

$$GOD(FADH_2) + M_{ox} \longrightarrow GOD(FAD) + M_{red} + 2H^+$$

$$M_{red} \xrightarrow{\text{anode}} M_{ox}$$

A variety of electron acceptors have been used as mediators in solution, for example, quinones, organic dyes, and ferricyanide [157]. Their use is limited, however, because they have to be added to the sample, and are toxic, pH-dependent, and often undergo autooxidation. Cass *et al.* (1984) used cyclic voltammetry to demonstrate that organometallic derivatives such as *ferrocene*, and its oxidized form, the *ferricinium* ion, are good mediators for glucose oxidase [158]. Ferrocene and its derivatives (dimethylferrocene, ferrocene carboxylic acid, etc.) have the advantage of being hydrophobic, which means that they can simply be adsorbed onto a carbon electrode. The potential necessary for their regeneration is low (160 mV/SCE) and hence the amount of electrochemical interference can be limited. An electrode that uses ferrocene as a mediator gives a current that is virtually independent of both pH and oxygen partial pressure. The range of the electrode response is quite large (1 to 30 mM of glucose) and the electrode has a good response time (60-90 s). These qualities led to the trial of such an electrode in monitoring a batch culture of *Escherichia coli* [159]. This electrode has, however, a number of weak points, notably its instability and short lifetime.

In order to ensure the electron transfer, the mediator must be present in both oxidized and reduced forms which must remain immobilized in the vicinity of the electrode. Unfortunately, the oxidized form of ferrocene, the ferricinium ion, is quite soluble and is released into the sample. The mediator should also have a rate constant for the reaction with glucose oxidase that is sufficiently competitive with that of oxygen, the natural mediator. This is true for ferrocene carboxylic acid, but the corresponding ferricinium ion is still too soluble.

The mediator can be made less soluble through the use of conducting polymers, which have properties that make them attractive matrices for redox enzymes. Conducting polymers are obtained by electropolymerization (see § 3.3.1.f), which is convenient in the construction of enzyme microelectrodes. A glucose-sensitive electrode can be prepared from a pyrrole solution containing glucose oxidase, which is polymerized by electrochemical oxidation in the absence of a mediator [46]. The presence of oxygen is necessary for the regeneration of FAD, which is essential for the enzymatic oxidation of glucose. In fact, it is impossible to obtain electron transfer directly from the active site to the platinum electrode using polypyrrole alone. This is because only the oxidized form of polypyrrole is conducting, the reduced form is insulating, and the redox potential of the polymer is very different from that of the cofactor.

The incorporation of the mediator in a polymer appears to be necessary to avoid the dependence on oxygen. This was achieved by Foulds and Lowe (1988), who attached ferrocene to the pyrrole before polymerization [45], and then by Hale *et al.* (1989) who attached it to a siloxane polymer [48]. In the latter, the half-life of the electrode obtained was 2 months.

(d) Direct transfer of electrons

Despite recent progress in the immobilization of mediators, they are normally avoided due to their toxicity and their possible release into the sample. Present research concentrates on creating an environment around the electrode that favors direct transfer of electrons from the enzyme to the electrode. As early as 1980, Kulys *et al.* [160] suggested the use of organic conducting salts to transfer electrons directly. They called these salts "organic metals" because their behavior at ambient temperatures was metallic, and not semiconducting like ferrocenes. Thus, an electrode that contains the salt NMP^+TCNQ^- (see § 3.3.1.f) can be used to oxidize glucose in an anaerobic environment at concentrations as low as 3 mM, with a potential of just 0.078 V.

A TTF^+TCNQ^- working electrode can also be used to determine glucose; the enzyme can be kept behind a membrane or simply adsorbed on the electrode surface. This electrode has been used to monitor glucose concentration in the brain continuously over 28 days [161]. Compared with other organic conducting salts, TTF^+TCNQ^- provides a high-performance glucose oxidase electrode, especially with regard to the background current, which is very weak [162]. Furthermore, this organic salt ensures a rapid rate of charge transfer to the enzyme so that the biosensor response is controlled only by the diffusion of the substrate in the membrane. The fact that the limiting factor is diffusion is an advantage because it depends on physical parameters that are reliable and relatively constant; the kinetic parameters of the reaction often depend on pH or the concentration of the active species. Moreover, when the process is diffusion-controlled, the amperometric biosensor response curve is linear with respect to substrate concentration because the flow of the substrate across the membrane is directly proportional to the concentration gradient (Fick's Law). In practice, the biosensor can be made to operate in the diffusion-controlled region by increasing the thickness of the membrane. This technique enlarges the range of detectable concentrations, but also increases the response time and reduces the sensitivity of the sensor. A compromise has to be adopted according to these requirements and to the kinetic parameters of the charge transfer to both the electrode and the enzyme.

Controversy remains over the mechanism of the action of organic conducting salts. This concerns whether they transfer the charge between the enzyme and the electrode by their conductivity, or whether, like ferrocene, they serve as mediators, with oxidized and reduced forms that relieve the cofactor and carry the charge from the enzyme to the electrode. Kulys (1986) supports the hypothesis of the existence of mediators [163], which, owing to mild dissolution of the organic salt, would be formed in the proximity of the electrode surface. These mediators would then come into contact with the active site of the enzyme, usually glucose oxidase, oxidize it, and then return to the electrode to be regenerated. Albery and Craston (1987) do not believe this mechanism [162] because they could not detect an appreciable concentration ($<10^{-6}$ M) of NMP^+TCNQ^-. The salt is too insoluble and the homogeneous kinetics are too slow to ensure effective electron transfer between the enzyme and the mediator. They are thus more in favor of a direct transfer of electrons, from the enzyme to the electrode, via an arrangement of successive positive and negative ions. This is made possible by the fact that the enzyme is strongly adsorbed on an electrode whose surface is made of organic conducting salts. Everything occurs in the heterogeneous phase, the substrate being the only mobile compound in solution. Whatever the mechanism, any dissolution, however weak, of organic conducting salts is undesirable in in-vivo applications; an implantable electrode has to be "clean" if it is to be used continuously.

4.5.2 Determination of polysaccharides

The determination of glucose required one enzyme, glucose oxidase, or perhaps glucose dehydrogenase [164], to convert the substrate into compounds that were detectable by the transducer. A number of enzymes are necessary for the detection of polysaccharides, first to hydrolyse them into glucose, and then to oxidize the glucose according to the techniques described above. Thus, the determination of sucrose [4] requires three enzymes: invertase, mutarotase, and glucose oxidase (GOD). These catalyse the following reactions:

$$\text{sucrose} + H_2O \xrightarrow{\text{invertase}} \alpha\text{-D-glucose} + \text{D-fructose}$$

$$\alpha\text{-D-glucose} \xrightarrow{\text{mutarotase}} \beta\text{-D-glucose}$$

$$\beta\text{-D-glucose} + O_2 + H_2O \xrightarrow{\text{GOD}} \text{gluconic acid} + H_2O_2$$

The determination of polysaccharides with enzyme electrodes is thus derived from the technique for determining glucose. The three enzymes are immobilized simultaneously (see § 3.3.1.d) on a membrane which is then attached to the tip of an electrode that is sensitive to oxygen (pO_2 electrode) or H_2O_2 (platinum electrode). In practice, any glucose electrode with an enzymatic membrane can be used, the glucose oxidase membrane is simply replaced with an invertase-mutarotase-glucose oxidase membrane. However, the presence of glucose oxidase means that this electrode is also sensitive to any glucose present in the sample and effectively measures the sum of the sucrose and glucose concentrations. A glucose-specific electrode must be used in addition to measure their separate concentrations; this electrode can be obtained by replacing the multienzymatic membrane with a membrane that only contains glucose oxidase. Another method is to transform all the sucrose into glucose via an enzymatic reaction, and determine glucose with a glucose oxidase electrode, before and after the transformation, thus yielding a measure of the sucrose concentration.

Lactose is determined using two immobilized enzymes: β-galactosidase and glucose oxidase [39]. The β-galactosidase catalyses the hydrolysis of lactose:

$$\text{lactose} \xrightarrow{\beta\text{-galactosidase}} \text{glucose} + \text{galactose}$$

The glucose produced undergoes enzymatic oxidation and is then determined with a platinum electrode, as described above.

Maltose is also determined through the coupling of two enzymes: glucoamylase and glucose oxidase [165, 166]:

$$\text{maltose} + H_2O \xrightarrow{\text{glucoamylase}} 2\ \text{D-glucose}$$

Higher polysaccharides, such as starch, cannot be determined with an enzymatic membrane covered with a dialysis membrane because the latter is only permeable to small molecules. Starch can only be detected by direct contact with an enzymatic membrane that contains α-amylase and glucose oxidase [166]. It is then hydrolysed according to the following reaction:

$$\text{starch} \xrightarrow{\alpha\text{-amylase}} n\ \text{glucose}$$

The determination of these various polysaccharides always involves the replacement of the glucose oxidase membrane of a glucose-sensitive electrode with the corresponding multienzymatic membrane.

4.5.3 Determination of alcohols

Two types of enzyme, alcohol dehydrogenases and alcohol oxidases, are used to determine alcohols such as ethanol or methanol. Alcohol oxidase has the advantage that it regenerates in the presence of oxygen, like glucose oxidase, but this requires a constant supply of oxygen for a reliable response. If the oxygen partial pressure is too low, then the calibration curve is only linear at low concentrations of alcohol [167].

In contrast, alcohol dehydrogenase (ADH) requires a coenzyme, nicotinamide adenine dinucleotide (NAD), whose reduced form (NADH) cannot be oxidized by oxygen. An electroactive species must act as a mediator to regenerate NAD. A variety of substances have been tested as mediators [168], phenazine methosulfate (PMS) rapidly oxidizes NADH but is unstable, and thioglycolate and methyl violet react too slowly. Flavin mononucleotide (FMN), however, is easily oxidized at the anode by a constant current, and NAD can be regenerated. Hence, ethanol is determined using the three following reactions:

$$CH_3CH_2OH + NAD^+ \longrightarrow CH_3CHO + NADH + H^+$$

$$NADH + FMN + H^+ \longrightarrow NAD^+ + FMNH_2$$

$$FMNH_2 \longrightarrow FMN + 2H^+ + 2e^-$$

The rate of NAD regeneration by an FMN mediator is still too slow to ensure a linear electrode response with respect to varying alcohol concentrations. Although the enzyme ADH eliminates the dependence of the electrode on oxygen, its use is limited because of the cost of the coenzyme NAD, especially considering that it has to be added to each sample measured. Further research has concerned the immobilization of NAD [42], or its electrochemical regeneration, with the aim of building an electrode without a mediator [169]. NAD is immobilized using cellulose membranes, which are modified by acetylation and then stretched to give low cut-off values (molecular weights of 500). The electrochemical regeneration of NAD uses a platinum electrode, which is polarized to 0.875 V/SCE, and not a vitreous carbon electrode which fouls up more rapidly in the presence of ADH.

The direct electron transfer from the coenzyme NADH to the electrode can be facilitated using organic conducting salts such as NMP^+TCNQ^- (see § 3.3.1.f). These salts reduce the cofactor redox potential required, which limits any interference phenomena and fouling up of the electrodes [43]. These problems can also be overcome using an electrode with a hydrophobic membrane (pO_2 electrode) on which alcohol oxidase and catalase [170] can be immobilized. These enzymes catalyse the following reactions:

$$RCH_2OH + O_2 \xrightarrow[\text{oxidase}]{\text{alcohol}} RCHO + H_2O_2$$

$$H_2O_2 \xrightarrow{\text{catalase}} H_2O + 1/2\ O_2$$

The catalase regenerates some of the consumed oxygen, and the response of the electrode is linear up to 25 mg/l. This electrode responds to ethanol and also to other alcohols, such as methanol, propanol, butanol and isopropanol.

Phenol is also determined using the oxygen electrode with phenol hydroxylase, which catalyses the following reaction [171]:

$$C_6H_5OH + NADPH + H^+ + O_2 \longrightarrow C_6H_4(OH)_2 + NADP^+ + H_2O$$

The determination of cholesterol is carried out using cholesterol oxidase coupled with a platinum electrode which detects the hydrogen peroxide formed [172].

4.5.4 Determination of lactate

The determination of lactate is important in medicine, because certain disorders are related to lactic acidosis, and in industry, for the control of milk and other food products. A lactate sensor can be constructed by coupling amperometric transducers with various enzymes, such as lactate dehydrogenase, cytochrome b_2, lactate oxidase and lactate monooxygenase. Lactate dehydrogenase requires NAD as a cofactor, whereas cytochrome b_2 uses potassium ferricyanide, which is much cheaper [173]. Cytochrome b_2 catalyses the following reaction:

$$CH_3CHOHCOO^- + 2[Fe(CN)_6]^{3-} \rightleftharpoons CH_3COCOO^- + 2[Fe(CN)_6]^{4-} + 2H^+$$

The ferrocyanide formed from the oxidation of lactate into pyruvate is electrochemically reoxidized into ferricyanide at an applied potential on a corrosion-proof electrode.

Lactate is generally determined by immobilizing lactate oxidase on an amperometric transducer, such as a pO_2 electrode [174, 175], to detect the oxygen consumption, or a platinum electrode [176], to detect the hydrogen peroxide formed. The enzyme catalyses the following reaction:

$$\text{L-lactate} + O_2 \xrightarrow[\text{oxidase}]{\text{lactate}} \text{pyruvate} + H_2O_2$$

The pyruvate formed is recycled into lactate using lactate dehydrogenase (LDH):

$$\text{pyruvate} + \text{NADH} + H^+ \xrightarrow{\text{LDH}} \text{L-lactate} + \text{NAD}^+$$

The coupling of these two enzymes amplifies the response of the lactate-sensitive electrode by up to 60 times, which gives a detection limit of 8×10^{-8} M lactate [174].

Lactate and pyruvate can be determined simultaneously in the same whole blood sample. Pyruvate is determined using an electrode constructed by immobilizing pyruvate oxidase on a platinum electrode to detect the H_2O_2 formed [177]. The L-lactate could also be transformed enzymatically into pyruvate, and then all the pyruvate in the sample reduced electrochemically by a mercury cathode polarized at –1.75 V [178].

4.5.5 Determination of amino acids

Amino acids may be determined potentiometrically by amino acid oxidases (L-AAO) (see § 4.4.3) because the corresponding enzymatic reaction liberates NH_4^+ ions:

$$RCH(NH_3^+)COO^- + H_2O + O_2 \xrightarrow{\text{L-AAO}} RCOCOO^- + NH_4^+ + H_2O_2$$

In a similar way to the reaction catalysed by glucose oxidase, this reaction consumes oxygen and liberates H_2O_2, and the oxygen-sensitive pO_2 electrode, or the H_2O_2-sensitive platinum electrode, can be used as transducers. If the pO_2 electrode is used, then it is preferable to immobilize catalase simultaneously with the glucose oxidase [3] so that the H_2O_2 produced is decomposed, and half of the oxygen can be re-used. Response curves for this electrode with various amino acids are given in Figure 4.29.

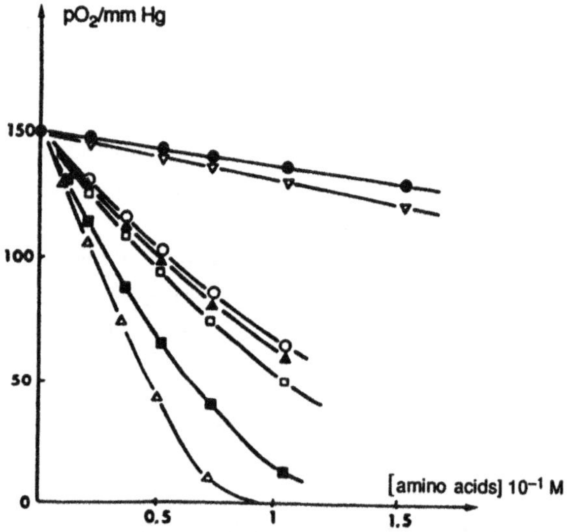

Figure 4.29 *Response of the L-AAO electrode, with a pO_2 transducer, to various amino acids:* ● *histidine;* ▽ *arginine;* ○ *methionine;* ▲ *phenylalanine;* □ *tryptophan;* ■ *leucine;* △ *tyrosine.*

The amino acid biosensor response is expressed in terms of the partial pressure of oxygen and varies with the nature of the amino acid concerned. Figure 4.29 shows the activity of the enzyme towards each amino acid and, consequently, demonstrates the non-specificity of the sensor. An electrode that uses L-lysine α-oxidase for the determination of L-lysine, for example, is also sensitive to L-ornithine, L-arginine, and L-phenylalanine [179]. Only electrodes that use decarboxylases give good specificity (see § 4.4.3.c).

4.5.6 Comparison of amperometric enzyme electrodes. Multienzyme electrodes

A large number of amperometric enzyme electrodes are described in the literature, and in Table 4.3 we only present some typical electrodes, together with their important characteristics [2-185]. Firstly, there has been an enormous research effort concerning the glucose electrode. Precise and rapid determination of glucose is essential in both clinical laboratories and process control in the fermentation and the food industry. Most research has concerned the problems of interference and oxygen dependence, which are solved through the use of mediators and conducting salts, rather than by improvements in the enzyme (glucose oxidase) or the electrodes (pO_2 or Pt). Analogous research is also underway on alcohol and lactic acid electrodes.

The determination of nitrogen derivatives requires the association of a number of enzymes that work in sequence or in competition. The amperometric determination of creatinin requires the coupling of an enzyme, creatininase, and a nitrifying bacteria [180], which oxidizes the ammonium ions formed into nitrite and then nitrate. The oxygen consumption of this oxidation can be monitored with a pO_2 electrode. The determination of creatinin is actually simpler using potentiometric techniques because a pNH_3 electrode covered with creatininase can detect the ammonium ions produced directly by exploiting the ammonium-ammonia equilibrium.

Bilirubin is determined using two enzymes which work *in sequence*. Hence, glucose oxidase produces H_2O_2 which, in turn, oxidizes bilirubin in the presence of peroxidase. In contrast, the determination of aminopyrine uses catalase and peroxidase, which work *in competition*. In the presence of aminopyrine, the quantity of oxygen produced by the decomposition of H_2O_2 by catalase is reduced because H_2O_2 is also a substrate for the oxidation of aminopyrine by peroxidase [181].

The two enzymes used to detect carnitine are carnitine dehydrogenase and diaphorase, which operate *in opposition*. The carnitine dehydrogenase catalyses the oxidation of carnitine by NAD^+, and the diaphorase regenerates the cofactor in the presence of ferricyanide [182].

Finally, the phosphate-specific electrode relies on the analysis of enzymatic inhibition. The phosphate ions inhibit one enzyme, alkaline phosphatase, thereby preventing the production of detectable glucose by a second enzyme, glucose oxidase [185].

Table 4.3(a) *Some examples of amperometric electrodes*

Analyte	Enzyme (E)	Immobilization technique	Transducer	Response time	Linearity	Reference
glucose	glucose oxidase	E + polyacrylamide	p O_2	30 s - 3 min	0 - 0.5 g / l	[2]
glucose	glucose oxidase	E + BSA + GA	p O_2	1 - 3 min	0 - 2×10^{-2} M	[3]
glucose	glucose oxidase	E / nylon	p O_2	26 - 38 s	1 - 20 mg / dl	[113]
glucose	glucose oxidase	E + polyacrylamide / Millipore	p O_2	1 - 2 min	0 - 200 mg / l	[142]
glucose	glucose oxidase	E + gelatine + GA	p O_2	2 - 3 min	0 - 2 g/l	[143]
glucose	glucose oxidase	E + polyacrylamide	Pt	1 min	0 - 1.5×10^{-2} M	[145]
glucose	glucose oxidase	E / collagen	Pt	2 - 3 min	10^{-7} - 2×10^{-3} M	[146]
glucose	glucose oxidase	E + BSA + GA	Pt	10 s	10^{-7} - 2×10^{-3} M	[147]
glucose	glucose oxidase	E + silane + GA	Sn O_2	10 min	10^{-4} - 10^{-2} M	[148]

Table 4.3(b)

glucose	glucose oxidase	E + p-benzoquinone + BSA/cellulose	Pt	0.5 - 3 min	10^{-6} - 3×10^{-3} M	[149]
glucose	glucose oxidase	E + silane + GA/silicone	Pt	2 - 3 min	0 - 6 g/l	[151]
glucose	glucose oxidase	E + acrylamide/dialysis membrane/ silicone membrane	Pt	1 - 10 min	0 - 2.5×10^{-2} M	[153]
glucose	glucose oxidase	E + BSA + GA	Pt	1 - 2 min	0 - 2.7×10^{-2} M	[154]
glucose	glucose oxidase	E / semipermeable membrane	Pt	6 min	10^{-5} - 7×10^{-3} M	[156]
glucose	glucose oxidase	E / cellophane	Pt	3 - 4 min	5×10^{-4} - 6×10^{-3} M	[157]
glucose	glucose oxidase	E + ferrocene graphite	graphite	1 - 1.5 min	10^{-2} - 3×10^{-2} M	[158]
glucose	glucose oxidase	E / NMP^+ $TCNQ^-$	Pt	-	0 - 2.6×10^{-3} M	[163]
glucose	glucose dehydrogenase	E + BSA + GA	p O_2	3 - 15 min	10^{-3} - 8×10^{-4} M	[164]
sucrose	invertase, mutarotase, glucose oxidase	E + collagen + GA	p O_2	2 - 6 min	0 - 10^{-3} M	[4]
maltose	glucoamylase glucose oxidase	E + collagen	Pt	6 - 7 min	3×10^{-7} - 5×10^{-4} M	[165]

Table 4.3(c)

starch	α -amylase glucose oxidase	E + BSA + GA	Pt	2 - 3 min	-	[166]
methanol	alcohol oxidase	E / dialysis membrane	Pt	2 min	0 - 3×10^{-3} M	[167]
ethanol	alcohol dehydrogenase	E + collagen	Pt	30 min	non-linear	[168]
ethanol	alcohol dehydrogenase	E / dialysis membrane	Pt	5 - 10 min	non-linear	[169]
ethanol	alcohol oxidase catalase	E + GA	p O_2	1 - 2 min	0 - 25 mg / l	[170]
phenol	phenolhydroxylase	E + polyacrylamide	p O_2	-	0 - 5×10^{-5} M	[171]
cholesterol	cholesterol oxidase	E + collagen	Pt	3 - 5 min	10^{-7} - 8×10^{-5} M	[172]
lactate	cytochrome b	E / cellophane	Pt	7 min	0 - 2×10^{-2} M	[139]
lactate	lactate dehydrogenase	E + collagen	Pt	30 min	non-linear	[168]
lactate	lactate dehydrogenase	E / cellophane	Pt	1 min	10^{-4} - 10^{-2} M	[173]
lactate	lactate oxidase lactate dehydrogenase	E + polyester	p O_2	10 - 15 min	5×10^{-9} - 5×10^{-7} M	[174]

Table 4.3(d)

lactate	lactate oxidase	E / activated membrane	Pt	2 min	2.5×10^{-7} - 2.5×10^{-4} M	[176]
amino acids	L - amino acid oxidase	E + BSA + GA	p O_2	1 - 3 min	10^{-4} - 10^{-2} M	[3]
lysine	lysine oxidase	E + gelatine + GA	p O_2	2 min	2×10^{-4} - 4×10^{-3} M	[179]
creatinin	creatininase nitrifying bacteria	E + GA / triamine membrane bacteria / cellulose	p O_2	< 3 min	5 - 100 mg / dl	[180]
bilirubin	glucose oxidase peroxidase	E + gelatine	Pt	3 - 5 min	0 - 5×10^{-5} M	[181]
carnitine	carnitine dehydrogenase diaphorase	E + semipermeable membrane	Pt	2 min	10^{-4} - 10^{-2} M	[182]
choline	choline oxidase	E / nylon	p O_2	1 - 2 min	2×10^{-5} - 2×10^{-4} M	[183]
H_2O_2	catalase	E + collagen	Pt	1.5 - 2 min	0 - 1.5×10^{-3} M	[184]
phosphate	alkaline phosphatase glucose oxidase	E + GA	Pt	4 - 8 min	0 - 10^{-3} M	[185]
ATP, NAD$^+$	hexokinase or GDH glucose oxidase	E + BSA + GA	Pt	2 - 3 min	-	[166]

4.6 Semiconductor enzyme sensors

Enzyme sensors involving semiconductors are called enzyme field-effect transistors, ENFET, and, as their name implies, exploit the association of an enzyme with a field-effect transistor (FET). The transistor has a metal oxide semiconductor field-effect transistor (MOSFET) structure, which is constructed from, for example, a p-type silicon substrate (Figure 4.30). This central channel is defined by placing two n-type semiconducting zones, called the source and the drain, on opposite sides of the substrate. A metallic gate is isolated from the channel by a thin insulating film (SiO_2), which also covers the upper face of the substrate.

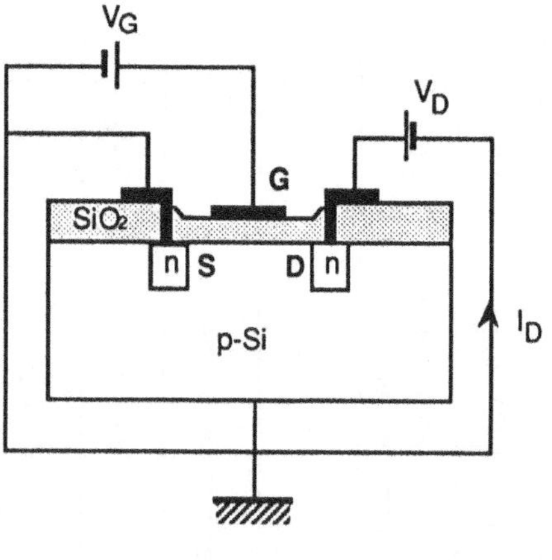

Figure 4.30 *Schematic representation of a MOSFET transistor (S - source; D - drain; G - metallic gate; Si - silicon substrate; SiO_2 - insulating layer).*

4.6.1 MOSFET sensors

In a MOSFET, the substrate and the source are both connected to the earth, and the gate-source and the drain-source potentials are given by V_G and V_D, respectively. When the transistor is not in a saturated mode, the drain current I_D is given by the expression [186]:

$$I_D = \mu\, C_{ox} \frac{W}{L} [(V_G - V_T)\, V_D - \frac{1}{2} V_D^2]$$

where μ is the mobility of the electrons in the channel, C_{OX} is the capacity of the insulating oxide, W/L is the ratio of the width of the channel to its length, and V_T is the threshold voltage. When $V_G > V_T$, the channel behaves as an n-type semiconductor and gives rise to a drain current. In a saturated mode, the expression above becomes:

$$I_D = \frac{1}{2} \mu\, C_{ox} \frac{W}{L} (V_G - V_T)^2$$

The MOSFET structure may operate as a sensor provided that the current I_D is a function of the parameter under study. The simplest modification consists of separating the metallic gate, and covering the SiO_2 layer with a material that is sensitive to the analyte. When the transistor is designed for the measurement of gases (GASFET), the adsorbed gaseous molecules dissociate and the charge is transferred into the oxide layer. This produces a variation in the dielectric constant of this layer and a modification of the drain current.

4.6.2 ISFET sensors

An ISFET (ion-sensitive field-effect transistor) sensor determines ions in solution and is built by replacing the metallic gate with a reference electrode (Figure 4.31). The SiO_2 layer is covered with a thin film of a material that has a strong interaction with the ion to be detected. For example, if the ISFET is intended to detect H^+ ions, a thin film of Si_3N_4 is fixed on the SiO_2 oxide layer by vapor deposition [51]. The interaction between the ions and the thin film changes the properties of the insulator-electrolyte interface, which leads to a modification of the threshold voltage, and hence a change in the drain current. Ion-sensitive sensors have been constructed in this way. One important example is the pH-sensitive ISFET, which can be transformed into a sensor for dissolved acidic or basic gases (pCO_2 or pNH_3) by adding a gas-permeable hydrophobic membrane. The gases diffuse across the membrane and react with the electrolyte liberating H^+ ions that are detectable with the ISFET.

4.6.3 *ENFET sensors*

An ENFET sensor is constructed by placing an enzymatic membrane over the ion-selective membrane of an ISFET (Figure 4.31). An ENFET is effectively a miniature version of an enzyme electrode. The product of the enzymatic reaction chosen for measurement is detected by the ISFET.

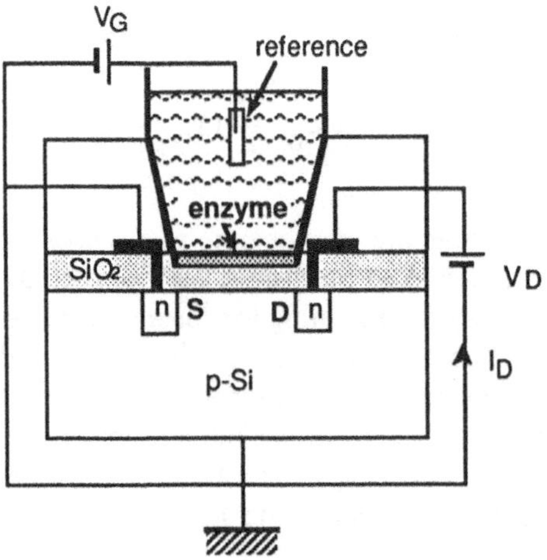

Figure 4.31 *Schematic representation of a semiconductor enzyme sensor (ENFET), in which the enzyme is immobilized on the sensitive component of an ISFET (S - source; D - drain; Si - silicon substrate; SiO_2 - insulator).*

Penicillinase is immobilized on a pH ISFET to obtain an ENFET that is sensitive to penicillin [7]. The variation in local pH, arising from the hydrolysis of penicillin into penicilloic acid, is measured by the adjacent ISFET. The response time of this ENFET varies from 30 to 50 seconds according to the thickness of the enzymatic membrane; its sensitivity is related to the buffering capacity of the solution. A differential measurement compensates for variations in pH or temperature, and a concentration range of 10^{-3} to 10^{-1} M penicillin can be obtained.

A urea-sensitive ENFET is constructed by immobilizing urease on a pH ISFET. The pH at the enzymatic layer varies in the presence of urea and the sensor obtained responds in 1 minute to urea in the concentration range of 5×10^{-5} to 10^{-2} g/ml [187, 188].

The pH-sensitive ISFET is also suitable as the basis of a glucose-sensitive ENFET. Glucose oxidase catalyses the transformation of glucose into gluconolactone, which, in turn, hydrolyses into gluconic acid which is detectable by the ISFET. The response of the sensor is linear between 10^{-4} and 10^{-3} M glucose [189]. The coimmobilization of glucose oxidase and gluconolactonase also helps by catalysing the hydrolysis of gluconolactone, and increasing the amplitude of the ENFET response in the presence of glucose [190].

Semiconductors of the type Ir-Pd-MOS are sensitive to gaseous ammonia, and can form the basis of a urea-sensitive enzyme sensor [191]. The corresponding enzyme, urease, is immobilized on the hydrophobic membrane of the transducer. The enzymatic reaction liberates NH_3, which diffuses through the membrane and is detected by the MOS transducer. The detection of gaseous NH_3 is much more selective than the measurement of pH. Only amines with low molecular weights are volatile enough to diffuse through the hydrophobic membrane by gaseous permeation and interfere in the response.

The pH-sensitive ISFET can also be used to determine hypoxanthine, an indicator of the freshness of fish [192]. Xanthine oxidase (XO) is immobilized on a polyvinylbutyral membrane over the insulating layer of SiO_2, and catalyses simultaneously two successive reactions:

$$\text{hypoxanthine} + O_2 \xrightarrow{XO} \text{xanthine} + H_2O_2$$

$$\text{xanthine} + O_2 \xrightarrow{XO} \text{uric acid} + H_2O_2$$

The ENFET thus detects hypoxanthine by detecting uric acid.

Cells from microorganisms can also be immobilized on a pH ISFET. A membrane containing cells from *Gluconobacter suboxydans* catalyses the conversion of ethanol into acetic acid, and can be coupled with a pH ISFET for the determination of ethanol [193].

Polyvinylbutyral membranes containing antibodies or receptors can also be immobilized on pH ISFETs [194]. The immobilization of the antibody anti-HSA (human serum albumin) on an ISFET gives an *immuno-FET* (IMFET) that is sensitive to HSA. The isoelectric point of HSA occurs at pH 4.8, and so the HSA molecule is negatively charged at pH 7, the pH at which the test is conducted. When the antigen-antibody coupling occurs, there is an accumulation of negative charge at the surface of the insulating layer, leading to a decrease in the potential of the sensitive component. The response increases with pH because the charge of the HSA protein rises progressively with pH. The charge is zero at the isoelectric pH.

An acetylcholine receptor may be immobilized in the place of the antibody on the ISFET. In the presence of the positively-charged acetylcholine, the potential difference between the sensing element and that of a *reference ISFET* (REFET) without a receptor changes in the same direction, indicating that the acetylcholine is strongly bound to its receptor. This is not the case for other proteins (catalase or BSA) when they are fixed in the place of the acetylcholine receptor.

Finally, the performance (response time and linearity) of these ISFET-based biosensors is presented in Table 4.4. This gives an overall view and comparisons can be drawn between the different sensors.

4.6.4 Comparison between ENFET sensors and potentiometric enzyme sensors

The sensitive component of an ISFET operates along the same principles as a potentiometric electrode. In both systems, there is no consumption of reaction product, and the interaction of ions with the sensitive component leads to a membrane potential. This potential is measured directly in the potentiometric electrode and indirectly in the field-effect transistor. The measurement of the activity of the ions is generally made at a constant current; the measured output potential is thus a function of the ionic activity of the solution. Once the ISFET and the potentiometric electrode are covered with enzymatic membranes to give an ENFET and an enzyme electrode, respectively, the two biosensors differ only in their method of transduction.

The potentiometric electrode measures the membrane potential directly and has a strong impedance, requiring a millivoltmeter with a very high input impedance ($Z > 10^{12}\ \Omega$). In an ISFET, the electrolyte

Table 4.4 *Some examples of ISFET-based sensors*

Analyte	Enzyme (E)	Immobilization technique	Response time	Linearity	Reference
penicillin	penicillinase	E + BSA + GA	30 - 50 s	10^{-4} - 2.5 x 10^{-2} M	[7]
urea	urease	E + GA / cellulose	1 min	5 x 10^{-5} - 10^{-2} g / ml	[187]
glucose	glucose oxidase	E + polymethacrylate	-	10^{-4} - 10^{-3} M	[189]
glucose	glucose oxidase gluconolactonase	E + polyvinylpyrrolidone (photolithography)	5 s	2 x 10^{-4} - 2 x 10^{-3} M	[190]
urea	urease	dialysis membrane / E / hydrophobic membrane	3 min	10^{-5} - 5 x 10^{-2} M	[191]
hypoxanthine	xanthine oxidase	E + polyvinylbutyral	4 min	2 x 10^{-5} - 10^{-2} M	[192]
ethanol	*Gluconobacter suboxydans*	alginate / nitrocellulose	10 min	0 - 20 mg / l	[193]
HSA	anti-HSA antibody	antibody + polyvinylbutyral	> 5 min	0.01 - 1 mg / ml	[194]
acetylcholine	acetylcholine receptor	receptor + polyvinylbutyral	2 - 4 min	10^{-7} - 10^{-5} M	[194]

and the internal reference of the potentiometric electrode are replaced by a solid-state contact with the sensitive component [195], which is integrated into the input stage of the millivoltmeter, hence the name *integrated sensor*. This device yields a low output impedance and a high signal to noise ratio. Miniaturization is possible using microelectronic techniques, and improves the cost and strength of these sensors. The absence of an internal reference solution means that the sensor can work under pressures up to 30 bars, and at temperatures up to 80 °C.

The most significant feature of an ENFET with an ISFET transducer is its small size, which means that the sensor is suitable for in-vivo biomedical determinations and in-vitro determinations in microsamples. The ENFET also has the advantage of only requiring a small amount of immobilized enzyme, but most potentiometric electrodes now opt for interchangeable and disposable enzymatic membranes so this advantage is only apparent. The response time is related to the thickness of the enzymatic layer, and is of the same order of magnitude in both ENFET and macro-electrodes, being about 1 minute when the enzyme is directly attached to the sensitive component, and about 3 minutes when a hydrophobic membrane is employed [7, 195].

The majority of ENFETs use a pH ISFET as a basic transducer, which is very sensitive to the buffering capacity of the medium and to interference. The working ISFET must therefore be associated with a reference ISFET (without a bioreceptor), termed a REFET, which also reduces the sensitivity of the biosensor to variation in pH or temperature [196].

4.7 Optical enzyme sensors

Optical enzyme sensors are sometimes called "optodes" or "optrodes" because of their similarity with electrodes. They are usually optochemical sensors exploiting the variation in the optical properties of the sample medium or an immobilized receptor. This variation is due to a chemical or physicochemical change. *Optical fibers* are convenient transducers for the construction of such sensors, provided that a material that responds optically to the sample can be immobilized on the tip of a fiber. The signal is then the intensity of the light reflected, scattered, or emitted. The association of fiber-optic technology with optical spectrometry allows the optical signal to travel long distances between the sample and the measuring device without being perturbed by electric or magnetic fields. Optical fibers are also very robust, which means that they can work in hostile or explosive environments, and that their miniaturization offers many interesting perspectives in medicine.

4.7.1 Principles of operation

The principal task of an optical fiber is to convey a light beam according to the optical path shown schematically in Figure 4.32. The optical fiber is made of a core surrounded by a cylindrical sheath.

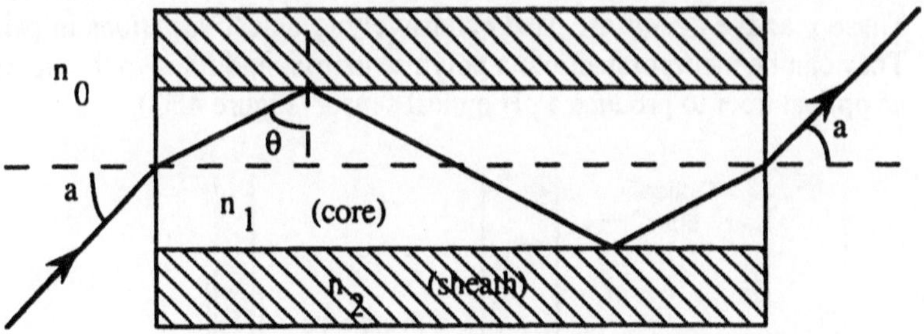

Figure 4.32 *Optical path of an optical fiber.*

The operation of an optical fiber is based on the principle of total reflection. For this to occur, the refractive index of the fiber n_1 must be higher than the refractive index of the external sheath n_2. The incident light may be transmitted along the fiber provided that it strikes the sheath at an angle which is larger than the critical reflection angle θ. The light is then totally reflected at this interface. The entrance cone for acceptable light has a half-angle a which is a function of the refractive indices of the core n_1, the sheath n_2, and air n_0:

$$\sin a = \frac{\sqrt{(n_1^2 - n_2^2)}}{n_0}$$

Fiber-optic sensors are divided into intrinsic and extrinsic sensors. *Intrinsic sensors* use the individual characteristics of the fiber. For example, interferometric sensors exploit the phase changes of the transmitted light. Such sensors can detect variations in pressure, temperature, and magnetic field. In contrast, *extrinsic sensors* use an external sensitive component. For example, the optical properties of an immobilized reactant may change upon interaction with the analyte. The fiber is thus used to transport the light to the sensitive component and then back to the measuring device.

Biosensors are usually based on extrinsic sensors. The biocatalyst is immobilized on the tip of an optical fiber, and the biosensor measures the absorption, fluorescence, or chemi/bioluminescence produced.

4.7.2 Optical sensors based on absorption

Absorption is the fundamental characteristic of colored indicators. These reactive agents are used extensively to detect variations in pH. They can be incorporated into a polymer matrix, and fixed to the tip of an optical fiber to produce a pH optical sensor (Figure 4.33).

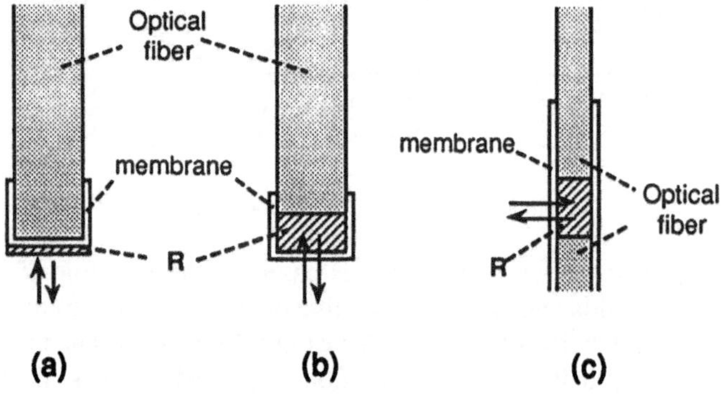

Figure 4.33 *Optical fibers using an absorbent reactant (R) fixed to the surface (a), or inside an external semi-permeable membrane (b,c).*

Indicators can be directly immobilized onto the tip of an optical fiber, or trapped between the fiber and a semi-permeable membrane. Configuration (a) in Figure 4.33 facilitates mass transfer and consequently decreases the response time of the sensor. Configuration (b), however, prevents interference from other substances because of the presence of the semi-permeable membrane. When gaseous compounds are determined, a hydrophobic membrane can prevent the access of ionic compounds to the reactive agent, which increases the sensor selectivity. The membrane can also be tubular (c). In this case, the diameter and thickness of the membrane must be as small as possible to facilitate the diffusion of the compound towards the reactive agent.

The immobilization of phenol red (phenolsulfonphthalein) in microspheres of polyacrylamide and polystyrene produces an optical sensor that is sensitive to 0.01 pH units in the pH range 7.0 to 7.4 [197]. Phenol red exists in two tautomeric forms, one acidic and one basic, which have different absorption spectra. Two light emitting diodes (LED) are used: a reference diode, emitting at 810 nm, and a working diode, emitting at 565 nm which corresponds to the wavelength of the maximal absorption of the basic form. This sensor can follow variations in pH between 6.5 and 9.0 [198]. Bromothymol blue has a maximal absorption at 593 nm and can be used instead of phenol red [199].

The basic form of p-nitrophenol has a maximal absorption at 404 nm, and can detect pH variations in the internal electrolyte solution of an optical ammonia sensor. The optical sensor is equipped with a hydrophobic teflon membrane to keep a solution of p-nitrophenol and ammonium chloride at the tip of the fiber; the sensor responds linearly to ammonia in the concentration range of 0.25 to 1 mM [200]. A colored indicator (oxazine perchlorate) can also be deposited on the wall of the optical fiber, in the form of a solid thin film. Using an LED that emits at 560 nm and a phototransistor, ammonia vapors can be detected down to 60 ppm, in the presence of 40 % humidity [201]. This low detection limit is possible due to multiple reflections (about 600) along the capillary tube.

Optical sensors that are sensitive to pH and ammonia can serve as a basis for the production of optical enzyme sensors [8]. Certain enzymes catalyse reactions with a variation in pH (penicillinase, cholinesterase, glucose oxidase or urease, etc.), or with liberation of ammonia (urease, glutaminase, amino acid oxidase) and are simply deposited onto the optical sensors.

Optical enzyme sensors do not necessarily require a colored immobilized reagent. Sometimes the product of the reaction is optically absorbent. One example is the simple optical sensor that uses immobilized alkaline phosphatase to catalyse the following reaction [9]:

$$\text{p-nitrophenylphosphate} + H_2O \xrightarrow[\text{phosphatase}]{\text{alkaline}} \text{p-nitrophenoxide} + \text{phosphate}$$

The product p-nitrophenoxide has a maximal absorption at 404 nm, and a selective filter for this wavelength is placed in front of the photomultiplier in order to determine p-nitrophenylphosphate.

Optical sensors based on the measurement of the absorption of light have many disadvantages. The use of colored indicators is not very convenient, especially as the conversion between the acidic and basic forms occurs in a restricted pH range reducing the range of detectable concentrations enormously. Furthermore, absorption techniques require monochromatic radiation and at least two optical fibers, one to convey the light beam to the sample and another for its return to the detector, because it is impossible to distinguish between the two beams on the basis of wavelength differences [202]. These separated or *bifurcated* fibers double the volume of the fibers.

The absorption technique gives a significant amount of light intensity, but requires a reference intensity. In spectrophotometers, a double beam is generally used, but a single wavelength serves as both the reference beam and the sample beam. The reference beam in optical sensors must have a wavelength at which neither the acidic nor the basic form of the indicator absorbs. In consequence, two LEDs are used, one of which, the reference, emits in the infrared to avoid absorption [198]. The use of a reference diode and supplementary optical fibers, to convey the reference signal to and from the sample, complicates the electronic circuitry considerably, and makes this system of measurement quite cumbersome.

4.7.3 *Optical sensors based on fluorescence*

Fluorescence is better adapted to analysis by optical sensors than absorption. Only a single fiber is used because the wavelength of emission is different from that of excitation. Furthermore, fluorescence is a sensitive technique that can detect very low concentrations. It has the further advantage of having a signal that is linear as a function of low concentrations.

(a) Direct measurement of fluorescence

A pH optical sensor can be constructed to measure the fluorescence emitted by fluoresceinamine immobilized on the tip of an optical fiber [10]. An argon laser excites the sample at 488 nm with a beam intensity of 1.5 W at the tip of the fiber. The response to pH is measured at 530 nm with a response time of 9 seconds. The sensitivity of this technique is poor due to difficulties in extracting the fiber from the mass of the polymer required for the immobilization of fluoresceinamine.

Other fluorescent colorants can be used to produce pH, pNH_3, and pCO_2 optical sensors. One example is the trisodium salt of 8-hydroxy 1,3,6-pyrenetrisulfonic acid (called HOPSA or HPTS) whose basic form can be excited at 470 nm to emit fluorescence at 510 nm [203, 204]. The CO_2 optical sensor requires a silicone membrane to contain a bicarbonate solution, just like potentiometric pCO_2 electrodes. A further similarity with potentiometric electrodes is that the pH, pNH_3 and pCO_2 optical sensors serve as the bases for enzyme sensors. The appropriate enzymatic membrane is simply attached. The optical sensor for ***penicillin*** uses penicillinase to catalyse the transformation of penicillin into penicilloic acid. The acid is detectable with a pH optical sensor that uses fluorescein isothiocyanate (FITC) as a colored indicator [205].

The ammonia optical sensor also uses an indicator with a fluorescent basic form. An optical biosensor for *glutamate* [206] is produced by immobilizing glutamate oxidase on this ammonia optical sensor. The enzyme catalyses the following reaction:

$$\text{glutamate} + O_2 \xrightarrow{\text{glutamate oxidase}} \alpha\text{-ketoglutarate} + H_2O_2 + NH_3$$

The determination of the ammonia produced gives a measure of the glutamate concentration present, which is possible between concentrations of 1 and 150 μM.

When a substrate is immobilized on the membrane in the place of the enzyme, then the sensor is capable of determining *enzymatic activity* [207]. Thus, an optical fiber equipped with an anion exchange membrane containing the colorant HPTS in the form of the ester of acetic or butyric acid measures the enzymatic activity of cholinesterases according to the following reaction:

$$\text{HPTS-acetate} \xrightarrow{\text{esterase}} \underset{\text{(fluorescent)}}{\text{HPTS}} + \text{acetic acid}$$

This method remains tentative because the substrate is consumed during the operation of the sensor, which makes calibration difficult and excludes any continuous measurement.

The creation of an optical sensor that detects the *cofactors* on which many enzymes depend is a much more attractive prospect. One example is NADH, which is associated with many enzymes, particularly dehydrogenases. This cofactor has a maximal absorption at 340 nm and a maximal fluorescence emission at 400 nm, which is easily detectable with a photomultiplier. NADH also has the advantage that it can be immobilized on the same support as the enzyme (see § 3.3.1.e). Fiber-optic biosensors based on the fluorimetric detection of NADH have been constructed for the determination of lactate and pyruvate [208]. These sensors use immobilized lactate dehydrogenase (LAD) to catalyse the following equilibrium reaction:

$$\text{lactate} + NAD^{+} \underset{}{\overset{\text{LDH}}{\rightleftharpoons}} \text{pyruvate} + \underset{\text{(fluorescent)}}{NADH} + H^{+}$$

This equilibrium is pH-controlled; a high pH value favors the formation of pyruvate. In order to determine lactate, the pH is fixed at 8.6, which facilitates the consumption of lactate and the production of fluorescent NADH. The intensity of the fluorescence increases with the lactate concentration, which can be measured from 10 to 60 μM.

Pyruvate is determined by setting the pH to a more acidic value (pH 7.4) to favor production of lactate. The NADH is then consumed, and the fluorescence diminishes with the pyruvate concentration, which can be measured from 0.2 to 1 μM. The presence of the equilibrium means that the response of the optical sensor is non-linear for both lactate and pyruvate.

The ethanol optical sensor also uses fluorimetric detection of NADH. Alcohol dehydrogenase (ADH) is fixed to a microporous teflon membrane in the presence of semicarbazide, which displaces the equilibrium by decomposing the aldehyde formed [206]:

$$CH_3CH_2OH + NAD^{+} \xrightarrow{\text{ADH}} CH_3CHO + NADH$$

$$CH_3CHO + H_2NNHCONH_2 \longrightarrow CH_3CH{=}NNHCONH_2$$

An excess of NAD^+ is maintained so that the rate of the enzymatic reaction depends only on the ethanol concentration. This gives a quasi-linear response between 1 and 9 mM of ethanol.

NADH can also be determined *in vivo* with an optical fiber using time-resolved fluorescence measurements [209]. This new technique exploits the rapid decrease in fluorescence over a few nanoseconds, and also eliminates the reduction in fluorescence by the optical fiber. The apparatus is shown in Figure 4.34. The excitation beam is provided by a sub-nanosecond nitrogen laser (337 nm). This excitation beam and the emitted fluorescence spectrum are transmitted by a single optical fiber with a diameter of 200 µm. If NADH concentrations can be directly measured *in vivo* then a wide variety of biosensors can be envisaged that exploit the enzymatic redox reactions involving this cofactor.

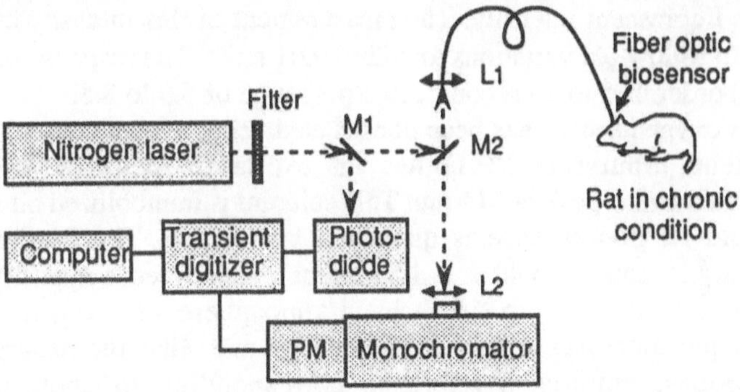

Figure 4.34 *Apparatus for time-resolved fluorescence measurements using a single fiber configuration (M - mirror; L - lens; PM - photomultiplier).*

(b) Measurement of fluorescence quenching

Fluorescence quenching occurs by a variety of mechanisms. Static quenching describes the situation when the interaction between the fluorescent molecule (F) and the quencher (Q) takes place in the ground state. Dynamic quenching occurs when they undergo a collisional process during the lifetime of the excited state of the fluorescent molecule, as shown below:

$$F + h\nu \longrightarrow F^* \text{ (excitation)}$$

$$F^* \longrightarrow F + h\nu' \text{ (fluorescence)}$$

$$F^* + Q \longrightarrow F + Q^* \text{ (quenching)}$$

A modulation of the fluorescent signal could also be due to inner filter effect and energy transfer effect, both of which require the fluorescent spectrum to overlap with the absorption spectrum of the absorber. The inner filter effect involves the absorption of the excitation and/or emission light by the quencher. In the energy transfer effect, the fluorescent molecule transfers its excited-state energy to an absorber which dissipates this energy through radiationless transition processes.

An optical pH sensor is produced by coimmobilizing a fluorescent compound, eosin, and an absorber, phenol red, on the tip of an optical fiber [210]. When the eosin is excited by an argon laser at 488 nm (i.e., in a region where the phenol red does not absorb), its emission occurs in a spectral region which overlaps with the absorption spectrum of the basic form of phenol red. This results in a transfer of energy from the donor (eosin) to the acceptor (phenol red). The quantity of energy transferred increases with increasing pH, leading to a reduction in the fluorescent intensity. The measurement of this intensity can be used to follow pH variations to within 0.01 units. The response time is of the order of 5 to 10 seconds, in a pH range of 5.5 to 8.5.

An oxygen sensor has been constructed using dynamic quenching. Perylene dibutyrate [211] has an excitation peak at 468 nm and an emission peak at 514 nm. This colorant is immobilized on silica gel and its fluorescence is quenched by oxygen. The fluorescent intensity is reduced to 40 % (with respect to its reference intensity) by oxygen in air, and to 16 % by 1 atmosphere of oxygen. This technique measures a transfer of energy and, like the absorption techniques, requires a reference corresponding to unquenched fluorescence. This reference uses derivatives such as bromonaphthalene which, when complexed with cyclodextrin, are not quenched by oxygen [212].

In the same way as the amperometric pO_2 electrode, the oxygen-sensitive optical sensor serves as a basis for the construction of enzyme biosensors. A glucose optical sensor is produced by immobilizing glucose oxidase in a polyacrylamide gel, which is held by a cellophane membrane on the tip of an optical fiber [213]. Naturally, the range of detectable concentrations depends on the initial partial pressure of oxygen. Concentrations down to 10 mM glucose can be measured in the presence of pure oxygen at atmospheric pressure.

When fluorescence is involved in the operation of an optical sensor, the emission time and the excitation/emission spectrum are both altered by the immobilization method (adsorption or covalent bonding), the concentration of the reactive species, and the nature of the solid support. It is therefore essential to opt for an immobilization method that does not reduce the fluorophore lifetime too much, otherwise the analyte may no longer be able to quench the fluorescence [214].

(c) Competitive bonding

Competitive bonding may occur between the analyte G and a fluorescent or absorbent ligand L for the same receptor site C, which is grafted on the tip of an optical fiber [215]. This interaction is governed by the equilibrium:

$$G + CL \rightleftarrows L + CG$$

The reversibility of this reaction must be preserved, and the optical detection of the free ligand L must be optimal. This imposes a number of conditions (Figure 4.35):

1. The ligand must be confined in a restricted space in front of the entrance cone of the fiber.
2. The receptor site must be outside the entrance cone so that the complex CL is not detectable.
3. The analyte must have easy access to the receptor site.
4. The reaction must rapidly reach equilibrium.

Schultz proposed [215] dextrane as a competitive ligand in the production of a glucose-sensitive optical sensor. This polyoside has a large molecular weight and is easily contained by a dialysis membrane. Glucose, in contrast, has a much smaller molecular weight and easily diffuses through the membrane to reach the receptor site, concanavalin A (Con-A). The dextrane is labelled with a fluorescent compound, fluorescein isothiocyanate (FITC), and so when it is not bonded to the Con-A outside the entrance cone, it is easy to locate by detection of its fluorescence. Glucose also has an affinity for Con-A, and so when glucose is present in solution, labelled dextrane is released into the solution in the entrance cone. This increases the amount of detectable fluorescence. Calibration relates the intensity of the emitted fluorescence to the concentration of glucose.

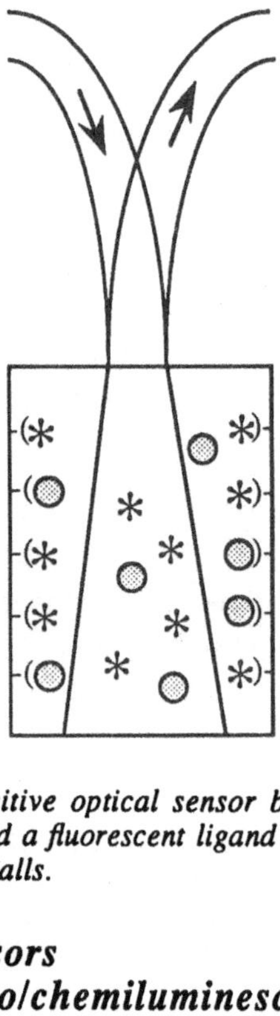

Figure 4.35 *Glucose-sensitive optical sensor based on the competition between glucose (●) and a fluorescent ligand (*) for concanavalin A attached to the detector walls.*

4.7.4 Optical sensors based on bio/chemiluminescence

The use of the emission of light from the reactions of some organic molecules is a significant contribution in the field of biosensors. This is especially important because the need for an excitation source is eliminated, and a single optical fiber is required to carry the light from the reaction to the detector. The measuring device is simpler (no excitation beam and no reference beam) and much more reliable.

(a) Bioluminescence

Bioluminescence is associated with the emission of light by living organisms such as the firefly, the glow-worm and various aquatic organisms. The production of light arises from a wide variety of sequences of biochemical reactions. A simplified explanation is that an enzymatic reaction is catalysed by *luciferase* and liberates a compound in its excited state, which emits lights as it falls back to its ground state level:

$$\text{luciferin} + O_2 + H_2O \xrightarrow{\text{luciferase}} \text{oxyluciferin}^*$$

$$\text{oxyluciferin}^* \longrightarrow \text{oxyluciferin} + h\nu$$

The radiation emitted by oxyluciferin* ($h\nu$) is similar to the fluorescence that can be produced by irradiating oxyluciferin via the usual method [216].

Luciferin includes a whole family of compounds whose heterocyclic structures vary from one organism to another. Most luciferase enzymes use the same cofactors as metabolic processes (ATP, FMN, NADH), and bioluminescence is easily associated with other types of biological reactions for analytical applications. Firefly luciferase is used to follow processes that use adenosine triphosphate (ATP) as a cofactor, for example, the measurement of biomass, the detection of a bacterial infection, antibiotic assays, and the monitoring of other enzymatic reactions that consume or produce ATP. Luciferase catalyses all these reactions according to the following overall reaction scheme:

$$\text{ATP} + \text{luciferin} + O_2 \xrightarrow[Mg^{2+}]{\text{luciferase}} \text{AMP} + \text{PP} + \text{oxyluciferin} + CO_2 + h\nu$$

The radiation is emitted at $\lambda = 562$ nm, and the excellent quantum yield (88 %) makes the technique very sensitive. The detection level for creatin kinase in the diagnosis of myocardial infarction is of the order of a femtomole (fmol = 10^{-15} mol). This enzyme is determined via the detection of ATP from the two following reactions:

$$\text{ADP} + \text{creatin phosphate} \xrightarrow[\text{kinase}]{\text{creatin}} \text{ATP} + \text{creatin}$$

$$\text{ATP} + \text{luciferin} + O_2 \xrightarrow{\text{luciferase}} \text{AMP} + \text{PP} + \text{oxyluciferin} + CO_2 + h\nu$$

Bacterial luciferase can be easily obtained from the culture of microorganisms (*Photobacterium phosphoreum, Vibrio harveyi*). Unlike firefly luciferase, this enzyme does not require luciferin. The luminescence comes from a complex formed from luciferase, $FMNH_2$ and fatty acid aldehyde. Moreover, commercialized luciferase contains NADH-FMN oxidoreductase, which means that the following reactions may occur:

$$NADH + FMN + H^+ \xrightarrow{\text{oxidoreductase}} NAD^+ + FMNH_2$$

$$FMNH_2 + RCHO + O_2 \xrightarrow{\text{luciferase}} FMN + RCOOH + H_2O + h\nu$$

The measurement of the emitted luminescence enables the detection of as little as 0.1 fmol of NADH, which means that traces of compounds whose enzymatic reactions consume or liberate NADH (or NADPH) can be detected. One example is the determination of ethanol using the reaction:

$$\text{ethanol} + NAD^+ \xrightarrow[\text{dehydrogenase}]{\text{alcohol}} \text{acetaldehyde} + NADH + H^+$$

The NADH formed is detected by oxidoreductase and luciferase, involving luminescent emission.

Other compounds are also determined in this way [217]: glucose (0.5-6 pmol); malate (20-250 pmol); and testosterone (0.2-5 nmol). The various enzymes used to determine these compounds can be immobilized on optical fibers, which transmit the bioluminescence to a photomultiplier.

One example is glutamate dehydrogenase (GDH), which is immobilized on the tip of an optical fiber in the presence of NADH-FMN oxidoreductase and luciferase. The GDH catalyses successively the following reactions [206]:

$$\text{glutamate} + NAD^+ \xrightarrow{\text{GDH}} \alpha\text{-ketoglutarate} + NH_4^+ + NADH$$

$$NADH + H^+ + FMN \xrightarrow{\text{oxidoreductase}} NAD^+ + FMNH_2$$

$$FMNH_2 + O_2 + RCHO \xrightarrow{\text{luciferase}} FMN + RCOOH + H_2O + h\nu$$

This optical sensor has a non-linear response with respect to glutamate concentration, but can measure concentrations up to 25 µM. Its response time is 1 to 2 minutes.

(b) Chemiluminescence

Chemiluminescence generally arises from chemical reactions in solution in which an oxidation occurs that involves either molecular oxygen or hydrogen peroxide (H_2O_2). The quantum yield in solution is generally very low because certain derivatives inhibit luminescence, notably oxygen. There are some exceptions such as luminol, acridinium salts, and some oxalic esters. Although the mechanism is poorly understood, the reaction of luminol (5-amino-2,3-dihydro-1,4-phthalazinedione) can be written schematically:

$$\text{luminol} + H_2O_2 + OH \xrightarrow{\text{peroxidase}} \text{aminophthalate} + N_2 + h\nu$$

$$(\lambda = 430 \text{ nm})$$

The best luminescence in aqueous solution is obtained in the presence of a catalyst and hydrogen peroxide. Chemiluminescence can be produced by using oxygen in a basic environment in the presence of dimethylsulfoxide (DMSO). The quantum yield of chemiluminescence is always low and never exceeds 1 %.

Freeman and Seitz [11] exploited the above reaction to produce a hydrogen peroxide-sensitive optical biosensor. Here, the luminol must be in excess with respect to H_2O_2 and the peroxidase must be immobilized on transparent polyacrylamide to ensure the best transmission of the chemiluminescence. A compromise must be adopted between the optimal pH of peroxidase, which produces a maximal amount of activated compound, and the optimal pH for the fluorescence occurring when the activated product arising from the luminol falls back to its ground state. The best compromise is achieved with a luminol concentration of 10^{-3} M and pH 9.

Could this sensor serve as a basis for a glucose-sensitive biosensor by covering it with a layer of glucose oxidase to catalyse the production of H_2O_2? Freeman and Seitz obtained results concerning luminol luminescence which show that the luminescence becomes very low below pH 9 where glucose oxidase is active. Thus, the glucose optical sensor [218] obtained would not perform as well as its amperometric counterpart. Nevertheless, chemiluminescence remains attractive because of its simplicity and rapid response time (4 seconds).

4.8 Thermal enzyme sensors

Thermal enzymes sensors, or *enthalpimetric* enzyme sensors, measure the concentration of a substrate using the variation in the enthalpy of an enzymatic reaction. According to whether the reaction is exothermic or endothermic, the increase or decrease in temperature of the reaction medium can be monitored using a temperature sensor. Thermistors or thermopiles are usually employed because of their large sensitivity to small variations in temperature. Two thermistors are placed in a thermally stable environment to allow a differential measurement of small changes in temperature. The method used to determine the variation in enthalpy depends on whether the enzyme is immobilized at the reactor or directly on the thermosensitive transducer.

When the enzyme is immobilized inside a reactor, usually an enzymatic column [219], two thermistors are placed on the outside of the column. One of the thermistors serves as a reference, and the other is the working thermistor; they can be placed in series or in parallel. When they are placed in series, only one enzymatic column is used, the reference thermistor is upstream from the column and the working thermistor is downstream from it. When they are mounted in parallel, the incoming flow of substrate is divided between two columns, one of which contains the immobilized enzyme. The working thermistor is placed downstream from the enzymatic column and the reference thermistor is placed downstream from the reference column. Although this second method is more complex and requires two pumps, it offers a better analytical performance because of the symmetry of the system.

The use of an enzymatic reactor associated with one or two thermistors on the outside of the column to measure the increase in temperature arising from the reaction, cannot be considered as a real biosensor. This is really microcalorimetry and requires cumbersome equipment (reactors, pumps, etc.), a large amount of immobilized enzyme and hence entails a large consumption of substrate.

A real thermal enzyme sensor must involve, as its name suggests, a thermal transducer (thermistors or thermopiles) with a layer of immobilized enzyme. The enzyme transforms the substrate and ensures the liberation or consumption of calories that are detectable *in situ* by the transducer.

4.8.1 Principles of thermal enzyme sensors

When an enzymatic reaction gives rise to a variation in enthalpy (ΔH) in the presence of substrate (S), then the change in temperature (θ) is a function of time (t) according to the expression [12]:

$$\frac{\partial \theta}{\partial t} = \frac{\Delta H}{\rho C} V_m \frac{[S]}{K_m + [S]}$$

where C is the thermal capacity, ρ is the specific gravity of the active layer, and V_m and K_m are the maximal rate and the enzyme Michaelis constant, respectively. This variation in temperature causes the diffusion of heat in the active layer covering the thermistor, which is governed by the equation:

$$\frac{\partial \theta}{\partial t} = \frac{\lambda}{\rho C} \frac{\partial^2 \theta}{\partial x^2}$$

where λ is the specific heat conductivity, and x is the distance from the point under study to the external surface of the membrane. The total variation in temperature is thus the sum of the kinetic and diffusion terms described above:

$$\frac{\partial \theta}{\partial t} = \frac{\lambda}{\rho C} \frac{\partial^2 \theta}{\partial x^2} + \frac{\Delta H}{\rho C} V_m \frac{[S]}{K_m + [S]}$$

When the thermal sensor response reaches the steady state then $\partial \theta / \partial t = 0$, and the following relationship applies:

$$\frac{\partial^2 \theta}{\partial x^2} + \frac{\Delta H}{\lambda} V_m \frac{[S]}{K_m + [S]} = 0$$

Numerical resolution of this equation gives the temperature profile at the interior of the enzymatic layer fixed to the thermistor. Theoretically it is now possible to find the temperature measured by the thermistor as a function of the different substrate concentrations in the sample medium, notably:

when $[S]/K_m \rightarrow 0$ (first order kinetics with respect to S), the temperature increases proportionally to the concentration of substrate S,

when $[S]/K_m \rightarrow \infty$ (zeroth order kinetics with respect to S), the temperature reaches a plateau whose height increases with the activity of the immobilized enzyme. This is limited by the maximal quantity of immobilized enzyme, and by the diffusion of the substrate in the active layer.

4.8.2 Construction of thermal enzyme sensors

In principle, a thermal enzyme sensor can be constructed simply by immobilizing a layer of the appropriate enzyme on a thermistor or a thermopile. In reality, the quantity of the heat liberated by some enzymatic reactions is so low that there is significant background noise from the thermal variation of the surroundings, particularly under stirring conditions. Two thermistors are used, one of which is a reference without any enzyme, and thus most of the thermal perturbations can be subtracted. The thermal perturbations due to convection can be further limited by surrounding the thermistors with a glass sheath with a few holes to allow the substrate access (Figure 4.36). During the determination, the two thermistors are immersed in the solution containing the analyte. If, for example, the reaction is exothermic, some of the calories produced diffuse towards the solution while the rest diffuse to the probe thermistor causing a variation in its resistance. The reference thermistor records the thermal evolution of the sample medium. A differential measurement indicates which part of the signal arose from the enzymatic reaction.

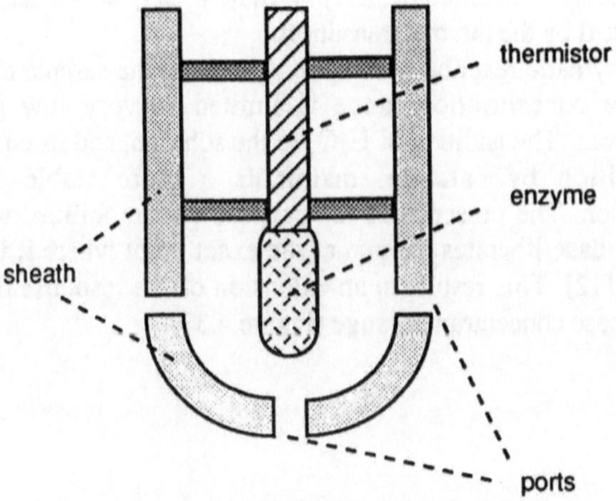

Figure 4.36 *Immobilized enzyme thermistor.*

(a) Determination of hydrogen peroxide

Catalase immobilized on a thermistor can catalyse the following reaction:

$$H_2O_2 \rightarrow H_2O + 1/2\, O_2 + \Delta H$$

The enzymatic activity is very high, and there is a strong variation in the enthalpy of reaction ($\Delta H = -23.44$ kcal/mol). The glass sheath in Figure 4.36 is therefore unnecessary. The catalase enzyme sensor responds linearly to H_2O_2 at concentrations between 10^{-2} and 5×10^{-1} M [12].

(b) Determination of glucose

Commercial glucose oxidase often contains catalase and the following reactions may occur:

$$\text{glucose} + O_2 \xrightarrow{\text{GOD}} \text{gluconic acid} + H_2O_2 + \Delta H_1$$

$$H_2O_2 \xrightarrow{\text{catalase}} 1/2\, O_2 + H_2O + \Delta H_2$$

The total variation in enthalpy, $\Delta H_3 = \Delta H_1 + \Delta H_2 \simeq 43$ kcal, is the value detected by the thermal transducer.

This enzymatic reaction uses up oxygen from the sample medium, and so the concentration range is limited to very low glucose concentrations. The addition of H_2O_2 to the solution, and its enzymatic decomposition by catalase, maintains a more stable oxygen concentration. The presence of the catalase coimmobilized with the glucose oxidase liberates oxygen at the exact point where it is being consumed [12]. This results in an extension of the response linearity for the glucose concentration range (Figure 4.37).

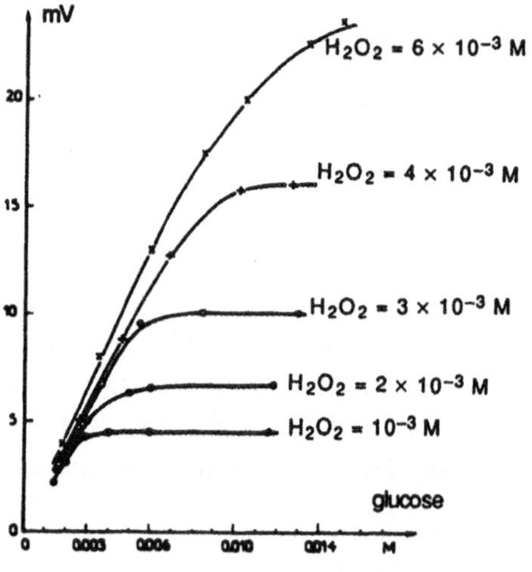

Figure 4.37 *Calibration curves for the glucose-sensitive bienzyme thermistor as a function of the concentration of hydrogen peroxide added to the sample medium.*

(c) Determination of urea

The following reaction can be catalysed by immobilized urease on a thermal transducer:

$$O{=}C(NH_2)_2 + H_2O \xrightarrow{\text{urease}} CO_2 + 2NH_3 + \Delta H \ (-13.77 \text{ kcal})$$

This reaction produces a low variation in enthalpy and so a glass sheath must be positioned around the thermistor to limit thermal diffusion. This device improves the sensitivity and the detection limit of the biosensor. The same pH effect is found for this sensor (Figure 4.38) as was found for other enzyme sensors, but the response time is much shorter (less than 10 seconds).

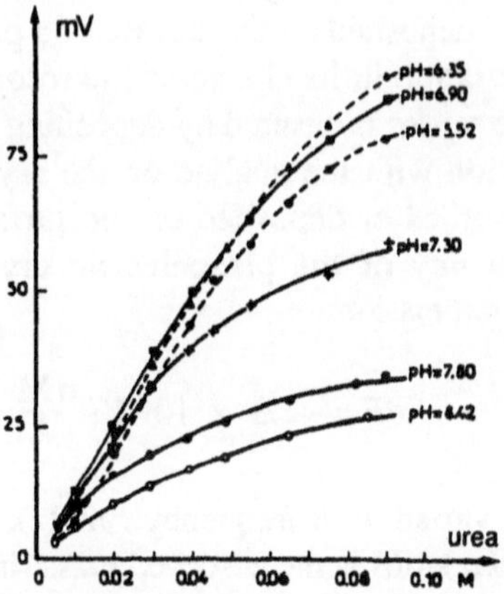

Figure 4.38 *Influence of pH on the response of the urease thermal sensor.*

Enzymatic thermistors also require control for any disturbance of the sample medium. The substrate of the enzymatic reaction must be added as regularly as possible to avoid high levels of thermal noise. The use of a laminar flow cell [220] can improve the detection limit for urea (10^{-4} M), but also reduces the sensitivity of the sensor.

Another method of improving the heat conductivity of the transducer is to deposit a drop of mercury on the thermistor to serve as a support for the immobilization of the enzyme [221]. This set-up is fragile and the response time is comparatively long (10-60 seconds).

Thermopiles may be used as transducers in the place of thermistors. A thermal glucose sensor has been constructed by immobilizing glucose oxidase and catalase on a set of 50 bismuth-antimony thermopile pairs [222]. This sensor can detect between 1 and 10 mM of glucose with a response time of 6 seconds. Thermopiles have the advantage of generating a passive signal which is practically insensitive to variations in the sample medium.

4.9 Piezoelectric enzyme sensors

4.9.1 Principles

The piezoelectric sensor is a gravimetric device which measures the mass of a sample deposited on the surface of a piezoelectric crystal by detecting the variation in its characteristic resonance frequency. The selectivity of the sensor is ensured by depositing a substance that has a specific interaction with the analyte on the crystal surface. When a substance is adsorbed or deposited on the surface of the sensor, the resonance frequency of the piezoelectric crystal diminishes [13] according to the expression:

$$dF = -2.3 \times 10^6 F^2 \frac{dM}{A}$$

where dF is the variation in frequency (in Hz), F is the fundamental frequency (in MHz), dM is the adsorbed mass (in g), and A is the area of the surface covered by the compound (in cm^2). Using this equation, it can be estimated that if the resonance frequency of a crystal is 9 MHz then its sensitivity is 400 Hz/μg. The sensitivity may reach 2600 Hz/μg for a crystal of 15 MHz, which gives a detection limit of 10^{-12} g [223].

A quartz crystal that resonates at 9 MHz is usually used. This comes in the form of discs, squares, or rectangles, which are about 15 mm wide and 0.15 mm thick. The surface of each face of the crystal is covered with metallic films between 0.3 and 1 mm thick, which serve as electrodes (Figure 4.39). This metal can be gold, silver, aluminum, or nickel. A thin film of the receptor is then deposited onto the surface of these two electrodes.

The piezoelectric sensor can be cumulative or reversible according to the type of interaction between the receptor and the compound it detects. The receptor can be a metal, a salt, or a protein (enzyme or immunoagent). AT-cut quartz crystal has temperature coefficients that are practically zero and is therefore well adapted for this mode of detection.

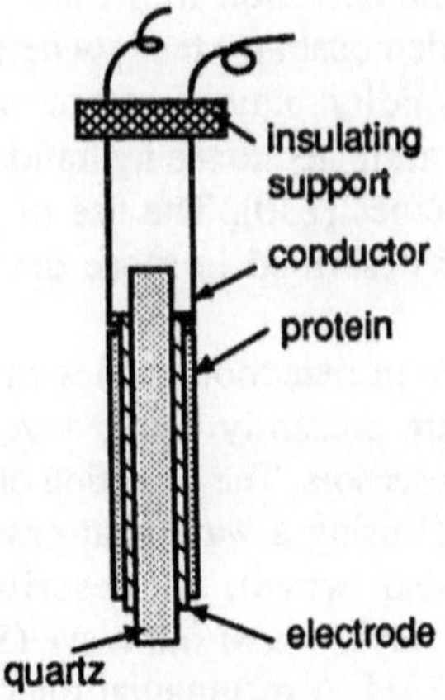

Figure 4.39 *A piezoelectric biosensor uses the vibration of a piezoelectric quartz disc and can detect the variation in mass due to coupling of the analyte with the protein.*

4.9.2 Construction of piezoelectric sensors

A quartz plate equipped with a gold electrode on each face can detect the adsorption of mercury vapor on the gold. A carbon monoxide sensor can therefore be constructed since this gas reacts with mercury oxide at 210 °C liberating mercury vapor [224] which is detectable with the piezoelectric sensor. Similarly, the deposition of copper complexes on a quartz crystal is used to detect organophosphates [225] such as diisopropylmethylphosphonate (DIMP), and modified cyclodextrins are used to detect benzene vapors [226]. The piezoelectric sensor is also used in the liquid phase to monitor the viscosity of fluids and gelation, and deduce, for example, the concentration of fibrinogen [227].

Enzymes can be deposited instead of inorganic compounds. The inhibition of cholinesterase by complexation with organophosphate derivatives is used for their detection in the gas phase [228, 229]. A variety of cholinesterases are used, for example, horse serum butyrylcholinesterase or acetylcholinesterase from the electric eel or erythrocytes. The enzyme is deposited in solution or immobilized with bovine albumin and glutaraldehyde, as described previously. The responses are linear between 4 ppb and 4 ppm of DIMP, and 5 and 50 ppb of malathion. The detection limits are 180 ng of DIMP and 4 ng of malathion. This demonstrates the low detection levels for these organophosphates. Humidity may interfere with the response of enzymatic quartz, and is attributed to the hydration of the enzyme in the presence of water molecules [230]. The use of piezoelectric sensors with antibodies will be described in more detail in the chapter on immunological sensors.

Aside from the different detection modes mentioned above, new measuring techniques are presently being developed to enlarge the scope of piezoelectric biosensors. The detection of the adsorbed analyte on a piezoelectric crystal using a wave that passes right through the crystal mass (bulk shear wave), as described above, may be supplemented using the *surface acoustic wave* (SAW) to transmit the vibration to the quartz [231]. A rectangular quartz plate is used, onto which the enzyme or immunoagent has been immobilized, and there are two sets of electrodes as indicated in Figure 4.40. Other piezoelectric materials can replace quartz. The SAW transducer works at higher oscillation frequencies and is thus able to detect lower masses.

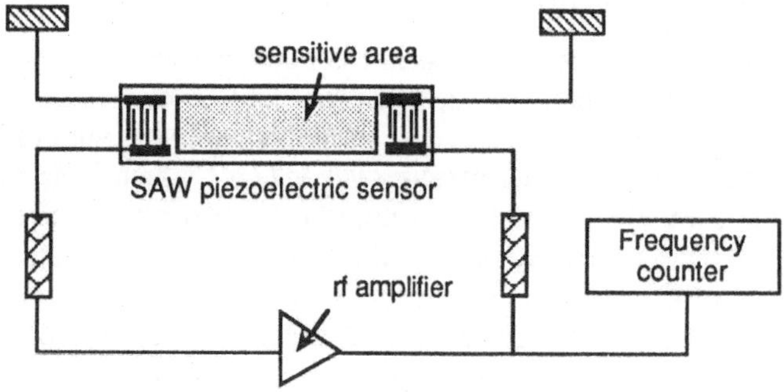

Figure 4.40 *Piezoelectric sensor using the surface acoustic wave.*

Finally, certain macromolecular materials have the property of swelling up under the action of parameters of a physical (temperature, electric field) or a chemical (pH, ionic strength, solvent) nature. This mechanical energy is exploited to produce an enzyme sensor based on the measurement of force. The enzymatic membrane is placed in contact with the cell that measures the force transmitted by the pressure of its swelling. This sensor has the advantages of being very sturdy, and having a catalytic element that is easy to change [232].

5

Microbial sensors

5.1 Preparation

Microbial sensors result from the combination of a microorganism with a transducer capable of detecting the metabolite involved. Microorganisms possess enzymatic systems that effect biological transformations. The immobilization of microorganisms on transducers is thus a logical step in the construction of biosensors. This approach is attractive because it eliminates extraction and purification processes, and because the enzyme is maintained in its natural environment thus avoiding problems like the regeneration of cofactors. Microorganisms are often immobilized by physical entrapment in gels, or using dialysis membranes, as described in § 3.3.2. The transducers employed in enzyme sensors are also used here. Potentiometric or amperometric electrodes (pCO_2, pNH_3, pO_2) are particularly useful because they already have a gas-permeable hydrophobic membrane (teflon or silicone). The microorganism is inserted between this membrane and a microporous or dialysis membrane. In contrast to enzymes, once the microorganism is immobilized it must be kept in a basal culture medium so as to maintain its activity.

5.2 Comparison of microbial sensors

Some examples of microbial sensors are given in Table 5.1 [233-241]. Ethanol may be determined using its oxidation by an acetic bacteria of the type *Acetobacter xylinum*, according to the following reaction:

$$CH_3CH_2OH + O_2 \longrightarrow CH_3COOH + H_2O$$

This oxidation is monitored with a pO_2 electrode whose response diminishes with ethanol concentration. Cellulose matrices treated with

Table 5.1(a) *Some examples of microbial sensors*

Analyte	Microorganism	Immobilization technique	Transducer	Response time	Linearity	Reference
ethanol	*Acetobacter xilynium*	cellulose material / GA	p O_2	2 min	0.1 - 3 mM	[14]
ethanol methanol	*Trichosporon brassicae*	adsorbed by suction on acetylcellulose membrane	p O_2	10 min	5 - 22.5 mg / l	[233]
phenol	*Trichosporon cutaneum*	dialysis membrane	p O_2	2 min	0.1 - 20 mg / l	[234]
cholesterol	*Nocardia erythropolis*	agar or polyacrylamide	p O_2	5 - 8 min	15 - 130 μM	[235]
glucose	*Escherichia coli*	agar	gold electrode	1 - 2 h	20 - 400 mM	[236]
glutamic acid	*Escherichia coli*	between silicone membrane and cellophane	p CO_2	5 min	100 - 800 mg / l	[237]
histidine	*Pseudomonas sp.*	dialysis membrane or Millipore GS membrane	p NH_3	7 - 12 min	10^{-4} - 3×10^{-3} M	[61]

Table 5.1(b)

arginine	*Streptococcus faecium*	dialysis membrane	p NH_3	20 min	5×10^{-5} - 1×10^{-3} M	[60]
L - aspartate	*Bacterium cadaveris*	dialysis membrane	p NH_3	5 min	3×10^{-4} - 7×10^{-3} M	[238]
L - cysteine	*Proteus morganii*	dialysis membrane	p H_2S	5 - 8 min	5×10^{-5} - 9×10^{-4} M	[239]
glutamine	*Sarcina flava*	dialysis membrane	p NH_3	5 min	10^{-4} - 10^{-2} M	[240]
serine	*Clostridium acidiurici*	dialysis membrane	p NH_3	3 - 5 min	1.8×10^{-4} - 1.6×10^{-2} M	[241]
nicotinic acid	*Lactobacillus arabinosus*	agar	pH	1 h	5×10^{-8} - 5×10^{-6} M	[59]
nystatin	*Saccharomyces cerevisiae*	collagen / GA	p O_2	3 - 25 min	0.5 - 70 units	[63]
BOD (biological oxygen demand)	*Clostridium butyricum*	collagen	p O_2	10 - 15 min	0 - 20 ppm	[62]

glutaraldehyde give natural bacterial films which are very thin [14] and which have a response time that is much shorter than that obtained by adsorbing *Trichosporon brassicae* on a porous acetylcellulose membrane [233].

Phenol is determined using a yeast, *Trichosporon cutaneum*, that oxidizes the phenol using two different enzymes which, in total, consume two oxygen molecules [234]:

$$\text{phenol} + O_2 + \text{NADPH} + H^+ \xrightarrow[\text{monooxygenase}]{\text{flavin}} \text{o-diphenol} + H_2O + \text{NADP}^+$$

$$\text{o-diphenol} + O_2 \xrightarrow{\text{dioxygenase}} \text{diene-adipic acid}$$

This is normal for microorganisms that contain a number of enzymatic systems. A pNH_3 electrode with immobilized *Streptococcus faecium* detects arginine through the presence of three enzymes [60], which catalyse successively the following reactions:

$$\text{L-arginine} \xrightarrow[\text{deaminase}]{\text{arginine}} \text{citrulline} + NH_3$$

$$\text{citrulline} + H_3PO_4 \xrightarrow[\text{transcarbamylase}]{\text{ornithine}} \text{ornithine} + \text{carbamoylphosphate}$$

$$\text{carbamoylphosphate} + \text{ADP} \xrightarrow[\text{kinase}]{\text{carbamate}} \text{carbamic acid} + \text{ATP}$$

$$\text{carbamic acid} \longrightarrow CO_2 + NH_3$$

This series of reactions increases the response time of the microbial sensor compared with that of the enzyme sensor for arginine, which only requires the presence of the first enzyme.

Walters *et al.* made a comparative study of the determination of histidine and published results concerning the bacterial *Pseudomonas sp.* electrode [61], and the ammonia-lyase enzyme electrode [125]; the ammonia lyase is purified from the same bacterial colony as

Pseudomonas sp.. The characteristics of the two sensors are given in Tables 4.2 and 5.1. The enzyme sensor has a better performance than the bacterial sensor, in terms of both response time and concentration range. The purification of the enzyme takes about a week, and the purified enzyme will keep for years without decomposing. This is not the case for the bacteria, which require periodic culture and harvesting. The bacterial sensor is, however, less sensitive to pH and temperature, but is the least selective since it is also sensitive to urea. Ammonia interferes with the response of both the enzyme and the bacterial electrodes, as they both use the same pNH_3 transducer.

Cholesterol is determined using an amperometric microbial electrode [235]. In contrast, glucose is determined using a potentiometric bacterial electrode, employing the reduction of lipoic acid in the presence of *Escherichia coli* [236]. The latter is based on bacterial growth, and the response time is very long, between 1 and 2 hours.

It is usually preferable to perform the conversion of a substrate by exploiting the enzymatic potential of a microorganism. The glutamate decarboxylase in *Escherichia coli* is used in conjunction with the pCO_2 electrode to measure glutamic acid [237], according to the following reaction:

$$\text{glutamic acid} \xrightarrow[\text{decarboxylase}]{\text{glutamate}} \gamma\text{-aminobutyric acid} + CO_2$$

Similarly, *Bacterium cadaveris* includes the enzyme L-aspartase (aspartate ammonia lyase). This microorganism can be fixed to a pNH_3 electrode and the enzyme catalyses the following reaction for the determination of aspartate [238]:

$$\text{L-aspartate} \xrightarrow{\text{L-aspartase}} \text{fumarate} + NH_3$$

Cysteine can be determined by a bacteria of the type *Proteus morganii*. The enzyme involved, cysteine desulfhydrase, catalyses the reaction:

$$\text{cysteine} \xrightarrow[\text{desulfhydrase}]{\text{cysteine}} \text{pyruvate} + NH_3 + H_2S$$

Here it is possible to use either a pNH_3 or a pH_2S transducer [239]. The pH_2S electrode is sensitive to CO_2, which limits its application somewhat because CO_2 is produced by the respiratory metabolism of bacteria. Consequently, the pNH_3 electrode is usually chosen for coupling with microorganisms for the measurement of amino acids [240, 241].

Microbial electrodes measure antibiotics in a very different way to enzyme electrodes. The penicillinase electrode transforms penicillin into penicilloic acid [108], which is detectable by the transducer. The microbial sensor for antibiotics makes use of the inhibitory effect of the antibiotic nystatin on the respiration of the yeast, *Streptomyces species* [63]. The enzyme sensor measures the number of moles of penicillin whereas the microbial sensor gives the unit activity of nystatin.

Finally, a microbial electrode in which the aerobic bacteria *Clostridium butyricum* is fixed to a pO_2 electrode [62] evaluates the biological oxygen demand (BOD) of waste water. This has the advantage of reproducing the natural phenomenon of the oxidation of biological compounds by microorganisms in the presence of oxygen.

5.3 Possibilities and limitations

In spite of some attractive applications, the construction of microbial sensors poses a number of problems. A sensor must be reliable and reproducible, and preservation of enzymatic stability is vital. Loss in enzymatic activity upon use of viable cells was considered to be counterbalanced by the growth of the biomass. In practice, this growth is difficult to control because it depends on a large number of physicochemical parameters. Furthermore, the biocatalytic matrix is not solid enough to contain the growth of the microbial population in the presence of its substrate. This results in a leak of cells and contamination of the sample medium. The microbial sensor is thus inappropriate for in-vivo measurements. Furthermore, in a complex biological environment, the large number of enzymes present in microorganisms makes any analytical interpretation very difficult.

The evolution of the biocatalytic matrix also complicates the modelling of such a system. The kinetics of microbial growth must be considered with the diffusion-reaction kinetics that we studied for enzymatic membranes. This extra factor modifies both the concentration of the biocatalyst and the diffusion coefficients of the different species because of dimensional constraints. This results in an increase in the response time of the sensor because this varies with both the thickness of the active layer and the diffusion coefficient of the substrate, which change during the operation of a microbial sensor.

Microbial sensors have not yet experienced the same expansion as enzyme sensors because of these various disadvantages. Nevertheless, they remain attractive in the fields where the use of enzyme electrodes is difficult (recycling of cofactors, evaluation of BOD, etc.).

6

Immunological sensors

Immunological sensors (or immunosensors) exploit the recognition involved in the coupling of an antigen with an antibody. This property can be associated with a variety of measuring techniques, which entail either the direct measurement of parameters resulting from the coupling, or the amplification of the coupling by enzymatic labelling.

6.1 Antigen-antibody coupling

Immunological sensors use immobilized immunoagents. An immobilized antigen is used to detect an antibody, or an immobilized antibody is used to detect an antigen. When an antibody interacts with an antigen, or vice versa, there is a variation in electric charge, mass or optical properties, which can be detected directly with a variety of transducers.

6.1.1 Electrochemical sensors

The difference in potential between an immunoreactive electrode and a reference electrode varies during the formation of an antigen-antibody complex. A titanium wire coated with anti-HCG (human chorionic gonadotropin), immersed in a solution containing 3.3 μg/ml HCG, gives a variation in potential of 3 mV [15]. This small potential difference is obtained with a high impedance ($10^9\ \Omega$) and the biosensor response time is longer than one hour. This method is therefore of limited practical use. The same device is also used to study enzyme-inhibitor coupling such as the trypsin-aprotinin reaction [19].

The antibody anti-dinitrophenol (anti-DNP) can also be measured directly by immobilizing DNP in a PVC matrix on a potentiometric electrode. The DNP is immobilized using an ion carrier [242], such as dibenzo-18-crown-6 with the ion K^+ (see § 3.3.3.b). This electrode is similar to ion-selective electrodes (ISE), especially those using liquid membranes containing ionophores, and also suffers from interference by mono- and divalent ions (e.g., Na^+, Ca^{2+}, Mg^{2+}).

The antibody anti-digoxin can also be determined potentiometrically using an electrode with a PVC membrane carrying the corresponding antigen coupled to an ionophore [21]. This electrode is immersed in a solution containing a constant concentration of a marker ion, K^+ (chosen for its compatibility with the ionophore portion of the conjugate) until a stable potential is reached. An antibody capable of fixing the antigenic portion of the conjugate is then added to the solution, and the variation in the resulting potential is proportional to the concentration of that antibody. The response of this electrode is not caused by the competition between the antibody and the K^+ ions for the conjugate sites, but by the reversible binding of the antibody to the antigenic portion at the membrane/solution interface. This hypothesis is confirmed by the fact that the electrode response at a given antibody concentration has a maximum for a K^+ concentration near 3×10^{-3} M. Under these conditions, the range of detectable antibody concentrations is between 1 and 20 μg/ml.

The ionophore is not essential if the antigen already has its own ionophoric character, which is true for digoxin, quinidine, and DNP. If one of these antigens is immobilized as such in a PVC matrix, then it shows the same response to its antibody as when it is in the presence of an ion carrier such as dibenzo-18-crown-6 [243].

Another potentiometric method of measuring antibodies, or their complements, exploits the property of antigens to lyse erythrocytes sensitized through their fixation to the corresponding antibodies. The blood cell ghosts are loaded with trimethylphenylammonium cations ($TMPA^+$), which act as markers. The liberation of these cations upon the action of the antigen is detected by an ion-selective electrode [244]. This method is impractical because the cations may leak out of the cells even in the absence of any antibodies, and there is a continuous and irreversible consumption of the marker. Other cations, such as tetrapentylammonium (TPA^+), can be imprisoned in liposomes [245]. These electrodes cannot be considered as real biosensors since the immunological reaction occurs in solution, and there is no immobilization of an immunoreceptor on the electrode.

6.1.2 Optical sensors

The antigen-antibody interaction can be optically detected in one of two ways. A variation in light intensity is produced by the coupling of an absorbent or fluorescent immunoagent, and can be measured using an optical fiber. Alternatively, the resonance of surface plasmons can be detected. The latter occurs when a light beam hits a metal-plated glass surface that is in direct contact with the antibody (or antigen). The measurement is made as a function of the angle of incidence of the light beam.

(a) Optical fibers

Optical fibers were discussed in § 4.7. They constitute useful supports and are excellent transmitters of light information from the point where the immunological coupling occurs back to the detector. An antibody can be immobilized on the tip of an optical fiber and, when coupling with a fluorescent antigen occurs, the fiber transmits the light to a photomultiplier. The metabolism of benzopyrene is monitored in this way, using the fluorescent product, benzopyrenetetraol (BPT), and its monoclonal antibody anti-BPT immobilized on an optical fiber [80]. The presence of the antibody permits the fixation, and, more importantly, the concentration, of the antigen at the entrance of the fiber, and a maximal amount of light is detected. The kinetics of this coupling is slow (12 hours to reach equilibrium) and the sensor must be calibrated at a constant incubation time. Using 15 minutes incubation, the antigen BPT can be detected down to 5×10^{-10} M.

The method of competitive fixation can also be exploited using immunoagents labelled with a fluorescent compound like fluorescein isothiocyanate (FITC). The antigen immunoglobulin G (IgG) is immobilized on the tip of an optical fiber and exposed to the antibody anti-IgG, of which a known quantity is labelled. There is then a competition between the labelled and unlabelled antibody molecules for the same antigenic site [246]. The fluorescence is only detectable from the labelled compound, and so the response of the optical biosensor is inversely proportional to the quantity of the (unlabelled) antibody corresponding to the analyte in the sample. In practice, the fluorescence can be excited using an argon laser and the sensor can be calibrated for a given incubation time. The detection limit varies according to the incubation time; the detection limit is 25 fmol for an incubation time of 20 minutes.

(b) Surface plasmon resonance

The detection of surface plasmon resonance (SPR) is one of the many techniques that exploit the properties of evanescent waves. This technique uses the intrinsic characteristics of the antigen-antibody complex, like refractive index and the thickness of an immunological layer deposited on a reflecting metal-coated glass surface. SPR involves determining the variation in reflectance (R) as a function of the angle of incidence.

For a particular value of the angle of incidence, i_r, a minimal reflectance peak is obtained corresponding to the SPR. The value of the angle i_r depends on the dielectric profile in the vicinity of the metallic layer. In order to determine an antibody (for example, anti-HSA (human serum albumin)), the corresponding antigen (an HSA antigen) is immobilized on the reflecting surface [247]. The light energy should be concentrated, as far as possible, in the proteinic layer, and the evanescent waves should not penetrate too far into the solution. Theoretically, for a resolution of $\Delta R/R = 10^{-3}$, a reasonable compromise is obtained with a silver layer 55 nm thick, onto which a 12 nm layer of a material with a high refractive index is deposited. The thickness of the adjacent protein layer can be measured down to 0.02 nm. For each antigen or antibody concentration measured, the reflectance R changes exponentially as a function of time, which means that direct calibration of the sensor is possible using the reaction half-life ($t_{1/2}$). This calibration curve is linear between 10^{-8} and 10^{-6} M of the antibody anti-HSA, giving a detection limit of 10^{-10} M [73]. This device remains, however, inconvenient compared with fiber optic sensors.

6.1.3 Semiconductor sensors

An antibody is an electrically charged protein, and its coupling with an antigen can give rise to variation in the dielectric constant which is measurable with a semiconductor sensor (see § 4.6.1). Thus, immuno-FET (or IMFET, immuno-field-effect transistors) are constructed by immobilizing immunoagents on field-effect transistors (FET).

The dissolution of the antigen cardiolipin in PVC followed by deposition of the solution on the metallic gate of the FET [248] produces a biosensor that responds to proteins. However, this sensor has a lack of selectivity towards serum containing the antibody

VDRL(+) and that without the antibody VDRL(–).

Other assays have concerned the immobilization of the antibody anti-alpha-fetoprotein on a silicon structure $Au/Si/SiO_2$ using glutaraldehyde. The interaction with the antigen in solution gives rise to variations in capacitance, which are measured at a potential difference of +4 V/SCE [249], and give a detection limit of 1 ng/ml. The method can also be applied to the detection of staphylococcus enterotoxin B by immobilizing the monoclonal antibody anti-enterotoxin (IgG) on the silica surface of the Si/SiO_2 heterostructure [250].

6.1.4 Piezoelectric sensors

Piezoelectric sensors detect variations in mass arising from the antigen-antibody coupling. The formation of the immunological complex increases the mass of the crystal thereby reducing its resonance frequency. Digoxin is determined by immobilizing the antibody anti-digoxin on the sensitive surface of a piezoelectric crystal, which is then placed in an environment containing the antigen for 30 minutes [231]. The antibody-antigen complex is dried and the measurement is made in the gaseous phase. The variation in frequency is linear with respect to digoxin concentrations between 100 and 800 ng/ml. This method is not very reproducible due to the presence of impurities such as traces of unevaporated water or solvent, which appreciably increase the mass of the immunological complex.

It is much more efficient to oscillate the piezoelectric crystal directly in the liquid phase. The liquid, however, absorbs a large proportion of the vibrational energy through its density and viscosity. This damping prevents the vibrational wave from penetrating the liquid mass any distance (estimated at a few micrometers). Kanazawa and Gordon [251] derived an expression for the variation of the frequency of a quartz crystal oscillating in a liquid medium:

$$\Delta f = -\, f_0^{3/2}\, (\eta\rho/\pi\mu_Q\rho_Q)^{1/2}$$

where f_0 is the frequency of the dry crystal, η and ρ are the absolute viscosity and the density of the liquid, respectively, and μ_Q and ρ_Q are the elastic modulus and the density of quartz, respectively. Glucose concentration in water can be directly measured using this method between 1 and 25 % by weight.

A single crystal face can also be exposed to a stationary or flowing liquid solution, and the antibody anti-IgG immobilized on this face using a thin film of polyacrylamide gel [252]. The variation in frequency is then monitored as a function of time, and falls each time a solution containing IgG is injected.

In order to determine IgG, the A protein is immobilized on the quartz crystal instead of the antibody anti-IgG [253]. In the presence of IgG, the resonance frequency of the crystal drops because of the fixation of IgG on the A protein. If the antibody anti-IgG is also added then this frequency drops further, which demonstrates that not all the fixation sites of IgG are saturated. Finally, rinsing with an acidic solution (pH 3) dissociates the IgG from the A protein, which returns the piezoelectric biosensor to a state where it is ready for a fresh IgG determination.

This method can also be applied for the detection of micro-organisms. The immobilization of the antibody anti-*Candida* enables the measurement of the pathogenic microbe *Candida albicans* at concentrations between 10^6 to 5×10^8 cells/cm^3 [254].

6.2 Enzymatic labelling

Labelling an antibody or antigen with an enzyme is an EIA or ELISA technique (see § 3.3.3.d). The presence of the enzyme ensures the consumption of substrate, or the generation of product, in a quantity large enough for detection by a transducer, hence the term *enzymatic amplification*.

6.2.1 Electrochemical methods

An electrode for the determination of vitamin H (biotin) is based on the competitive affinity between biotin and its homologue HABA for the same protein, avidin [84]. The catalase-labelled HABA-avidin complex is immobilized on a pO_2 electrode to produce a biotin-sensitive sensor [255]. Biotin dissociates some of the HABA-avidin complexes by its strong affinity for avidin (Figure 6.1), leading to a reduction in enzymatic activity at the pO_2 electrode which can be related to the concentration of biotin. The coupling between biotin and avidin occurs more from bioaffinity than from an immunological process.

Enzymatic labelling is generally used for immunological

determinations. The usual method exploits the *competition* between the labelled and unlabelled antigens for an antibody immobilized on an electrode.

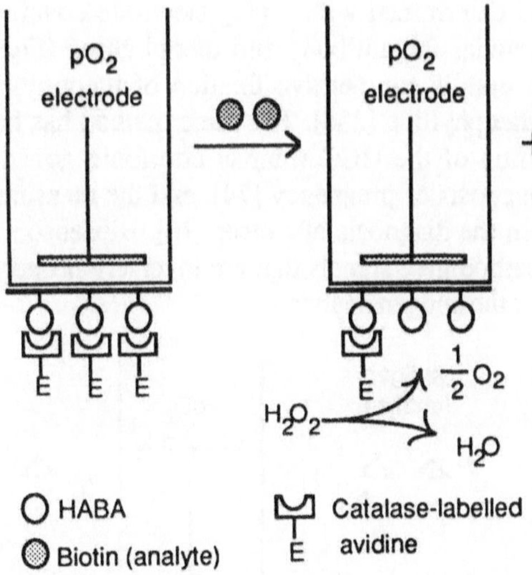

Figure 6.1 *Scheme showing the principle of a bioaffinity sensor. The pO_2 transducer evaluates the enzymatic activity by the amount of product (O_2) formed.*

Firstly, the antibody is immobilized on a transducer capable of following the reaction of the enzyme that labels the antigen. The biosensor is then placed in a medium containing a known quantity of the enzyme-labelled antigen and an unknown concentration of the unlabelled antigen analyte. There is then a competition for the antibody, and when the quantity of the labelled antigen fixed to the biosensor is low then the concentration of unlabelled antigen is high. After rinsing, the quantity of enzyme-labelled antigen is revealed by injection of the substrate of the enzymatic reaction. During this step, the immunosensor is simply playing the role of an enzyme electrode whose response is proportional to the activity of the fixed enzyme. Finally, the biosensor is regenerated by washing with a buffer solution of glycine-HCl to dissociate the antigen-antibody complex.

This method has been used to construct an immunological sensor for albumin using the antibody anti-HSA (human serum albumin). The albumin is labelled with catalase whose activity is monitored using a pO_2 electrode [77]. The pO_2 electrode can also be covered with a membrane containing the antibody anti-insulin, and determine insulin by competition with glucose oxidase-labelled insulin.

Theophylline is determined with a pO_2 electrode covered with a membrane that includes the antibody anti-theophylline (Figure 6.2). The measurement entails competitive fixation of theophylline with catalase-labelled theophylline [256]. The same method has been used for the determination of the HCG (human chorionic gonadotropin) hormone in the diagnosis of pregnancy [74], and the measurement of alpha-fetoprotein in the diagnosis of cancer [76]. Biosensors that use the competition method give signals that are inversely proportional to the concentration of the antigen analyte.

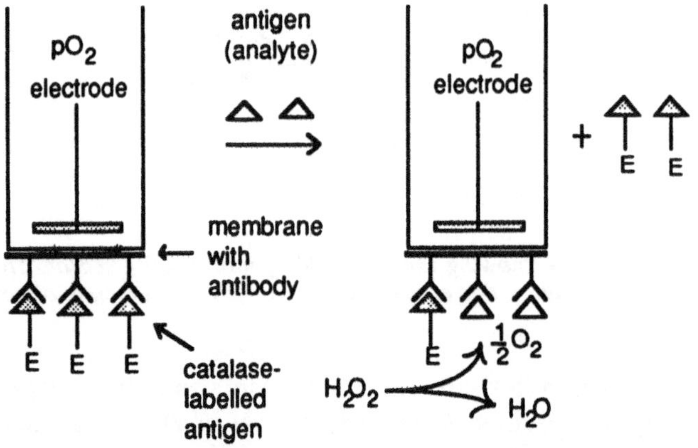

Figure 6.2 *Principle of the operation of an immunological sensor using the competition method.*

Antigens can also be determined indirectly, provided that their concentrations are low enough not to saturate the antibody immobilized on the electrode. Once the antigen is fixed to the immobilized antibody, an enzyme-labelled antibody is used to bind the antigen, and the enzymatic measurement is made (Figure 6.3). The antigen is thus determined *sandwiched* between the immobilized antibody and the enzyme-labelled antibody. This method has been attempted for the determination of the surface antigen of hepatitis B. However, the method is lengthy as it requires two coupling phases, 30 minutes to fix the antigen to the antibody IgG, and then 2 hours to fix the peroxidase-labelled IgG to the antigen [257].

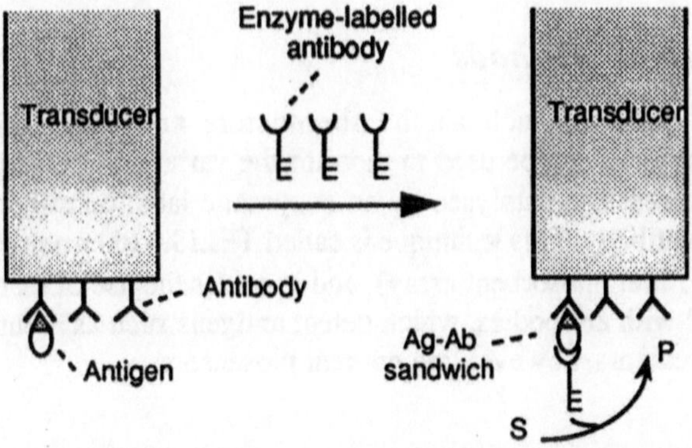

Figure 6.3 *Immunological sensor based on sandwich immunoassay. The antigen concentration is given by the amount of the labelled antigen-antibody (Ag-Ab) complex. The concentration of the immobilized enzyme E is measured by the enzymatic decomposition of the substrate S into a product P that is detectable by the transducer.*

The antibody IgG can also be directly immobilized on a vitreous carbon electrode. It then undergoes coupling with, firstly, the antigen analyte and, secondly, glucose oxidase-labelled IgG antibody [258]. This approach brings the catalytic and electrochemical sites close together, thereby reducing the biosensor detection limit. Immunological sensors that use the sandwich immunoassay method give signals which are directly proportional to the concentration of the antigen analyte.

The amperometric (pO_2) electrodes described above follow the *activity of the enzymes* used in immunological labelling; potentiometric electrodes, such as the pNH_3 electrode, may also be used. These detect the activity of urease [6] for the immuno-electrochemical measurement of BSA (bovine serum albumin) and cyclic AMP, or the activity of asparaginase for the measurement of cortisol and dinitrophenol [259].

6.2.2 *Optical methods*

Labelling with an enzyme that catalyses chemiluminescent reactions gives rise to signals that can be detected optically with the help of an optical fiber. For example, insulin can be determined by monitoring the activity of peroxidase immobilized on the antibody anti-insulin with the substrate luminol-H_2O_2 [260].

6.2.3 Thermal methods

Temperature sensors, such as the thermistors and thermopiles described in § 4.4.4, can be used to monitor the variation in enthalpy arising from a reaction catalysed by an enzymatic label on either an antigen or an antibody. This technique is called TELISA (thermometric enzyme linked immunosorbent assay), and involves the use of reactor columns lined with antibodies, which detect antigens such as albumin [261]. These reactors, however, are not real biosensors.

7

Other biosensors

Substances that are neither enzymes, microorganisms nor immunoagents can also be used in conjunction with transducers to construct biosensors. These include sections of animal and plant tissue, neuroreceptors and chemoreceptors.

7.1 Animal tissue

The use of whole cells containing biocatalysts is not limited to bacteria and other microorganisms. Fragments of animal tissue sometimes have high enzymatic activity which may be exploited. A section of pig surrenal cortex fixed to a pNH_3 electrode performs the selective determination of glutamine [119] with little interference from amino acids and more stability than an electrode using glutaminase. Similarly, samples of rabbit muscle tissue immobilized on a pNH_3 electrode can determine adenosine 5'-monophosphate (AMP) in a concentration range of 10^{-4} M to 10^{-2} M. This sensor has a response time of 2-8 minutes, which increases to 12-30 minutes after 4 days [262].

Animal tissue maintains the enzymes in an environment optimized by nature, in the same way as microorganisms. The structural features of the samples also provide useful mechanical properties for the production of biosensors.

7.2 Plant tissue

Another class of bioselective membranes uses leaves or vegetable tissue in conjunction with gas-sensitive electrodes. For example, cucumber leaves are fixed to a pNH_3 electrode for the potentiometric determination of cysteine [263]. The construction of this sensor entails removing the cuticle from the epidermic layer of the leaf, which is then orientated towards the sample; the other face is placed in contact with the hydrophobic membrane of the gas-sensitive electrode. The entire set-up is covered with a dialysis membrane. The presence of the enzyme desulfhydrase transforms L-cysteine into pyruvate and NH_3, and the ammonia produced is detected by the electrode.

Other biocatalytic systems have been found in fruit and vegetables. A "bananatrode" was constructed for the determination of dopamine by exploiting the presence of phenol oxidase in banana tissue. The ascorbic acid oxidase in cabbage or squash can be exploited to determine ascorbic acid [264], and the tyrosine oxidase in sugar beet can determine tyrosine [87] using the oxygen electrode. It is even possible to use a fragment of a cabbage leaf fixed to a pNH_3 electrode for the determination of sulfoxides [264].

The response time of these biosensors is improved by mixing the plant tissue with a carbon paste [90], when 95 % of the response can be reached in just 12 seconds. The measurement is made by flow injection analysis (FIA) and a detection limit of 1.3×10^{-6} M dopamine is possible for a biosensor that uses banana tissue in carbon paste. The response of this biosensor is linear to 9×10^{-5} M dopamine.

7.3 Other receptors

Chemoreceptors are important bioanalytical tools because of their specific abilities in molecular recognition. Associated with a suitable receptor, a biosensor can detect proteins, toxins, hormones, chemical derivatives, or living cells, in both gaseous or liquid media. The weak dissociation constant of the complex formed ($K \approx 10^{-9}$ for the cobratoxin/acetylcholine-receptor complex, and $K \approx 10^{-10}$ for the insulin/insulin-receptor complex) means that such biosensors can be purified by affinity chromatography, and detect a specific ligand associated with the bioreceptor at extremely low concentrations.

A recent sensor exploits a modification in the *dielectric constants* of the biochemical layer and the aqueous film resulting from the association of the receptor with its ligand. This interaction can be observed with a planar capacitive transducer [93] equipped with a silicon substrate covered with an SiO_2 layer 1 μm thick. The protein $Ca^{2+}Mg^{2+}$ ATPase can be immobilized on this layer to detect the snake

8

Use and application of biosensors

8.1 Operating methods

Once a biosensor has been conceived and constructed, it can be installed in a number of different ways according to the objective required. It may be simply immersed in the sample medium, placed in a cell into which a flowing sample is injected, or integrated into an automated system.

8.1.1 Direct determination in the sample medium. The batch method

Biosensors are simple in concept and have a selective character. They are thus usually capable of determining a compound directly in a sample medium (the batch method). Sometimes, it is desirable to place the sample in a measuring cell that allows the normal working conditions of the biosensor to be achieved (especially pH and temperature). Figure 8.1 shows an example of direct measurement in a batch system using an enzyme electrode and its associated analytical apparatus.

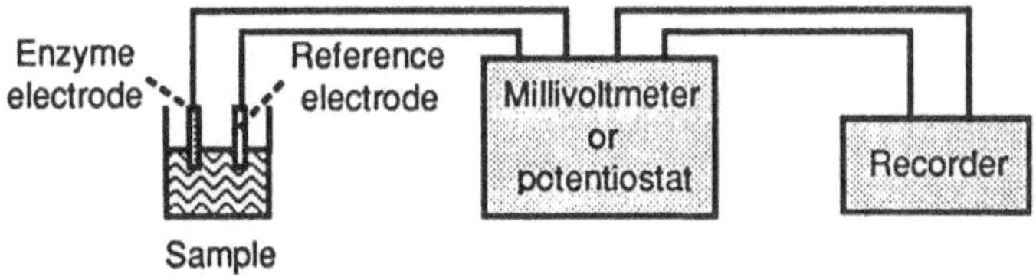

Figure 8.1 *Direct determination using an enzyme sensor in a batch system.*

venom Mojave from the snake *Crotalus scutulatus scutulatus*. Alternatively, an acetylcholine receptor can be immobilized on the SiO_2 layer for the specific determination of acetylcholine; no response is observed in the presence of other neurotransmitters such as glutamate, GABA, norepinephrine, histamine, 5-hydroxytryptamine, or dopamine. Furthermore, if acetylcholinesterase (AChE) is coimmobilized with the acetylcholine receptor then the biosensor responds to derivatives that inhibit AChE, such as organophosphates or carbamates.

Other biosensors use the chemoreceptor structure directly. One example is the antennules of the blue crab *Callinectes sapidus*, which form its chemosensitive organ (olfactive and gustative) (see § 3.3.4.b). This biosensor is sensitive to a variety of amino acids [92], especially glutamic acid, which is unsurprising since this acid is important in the composition of food proteins. This biosensor differs from other biosensors in that the antennule plays the role of both bioreceptor and transducer, and thus has a response time of a few milliseconds. The range of detectable concentrations can be very large (6 orders of magnitude) and the detection limit is very low (10^{-15} M). However, the average lifetime of this biosensor is about 12 hours, which limits its application considerably. The sensor is also susceptible to adaptation phenomena because once the system is saturated it stops responding to variations in concentration and has to be completely rinsed to restore its original activity.

Future olfactive (olfactive mucus) detection systems will probably be ultrasensitive chemical sensors [16] detecting nucleotides and hormones (e.g., pheromones). Biosensor research may also become orientated towards new bioreceptors [265], and perhaps even new biocatalysts, such as artificial enzymes [266].

A reference electrode, such as the standard calomel electrode (SCE), is placed next to the enzyme electrode. The reference electrode may be combined with the working electrode, as is the case in the pH-sensitive glass electrode. The electrodes are connected to a millivoltmeter for potentiometric measurements, or to a potentiostat for amperometric measurements. The system is connected to a recorder which monitors the biosensor stability and the progression of its response curve towards a steady state. This recorder may be replaced with a data acquisition unit which also gives the slopes of the response curves and relates them directly to the analyte concentration thereby increasing the rate of measurement.

In this method, the volume of the sample may vary a great deal, with a minimum of a few microlitres. If the volume is too small, however, the consumption of the substrate by the enzyme biosensor may result in a drift in its response. Direct measurement in the sample medium is generally only used during biosensor design to study the response time and plot the calibration curve. The sample medium is changed simply by rinsing the sensor. The method has the disadvantages of not providing a continuous measurement and being limited to manual use.

8.1.2 Flow injection analysis (FIA)

When the analyte is in a flowing liquid then flow injection analysis (FIA) can be used. The compound may be separated from any interfering elements beforehand, according to the specificity of the detector. This separation is normally performed using chromatographic techniques, such as high pressure liquid chromatography (HPLC). However, when the biosensor has a high selectivity this separation is unnecessary. If the sample volume is very large then the compound may be directly determined as a flowing liquid. Alternatively, the sample may be injected in small quantities directly into a continuous flow of carrier liquid without air segmentation; this is the true principle of FIA [267]. Air bubbles are not introduced into the carrier stream in FIA, although they were previously thought to be necessary for mixing the reactants (a technique used in Technicon biomedical apparatus). The liquid sample is introduced through an injection valve with a loop with an accurately known volume and is transported to the biosensor.

(a) Determination of substrates

The determination of the substrate of an enzymatic reaction by a biosensor associated with FIA is simple because the addition of a reactive species is unnecessary. The substrate is simply injected into the carrier stream using an injection valve (Figure 8.2). The substrate is detected and determined when it comes into contact with the biosensor.

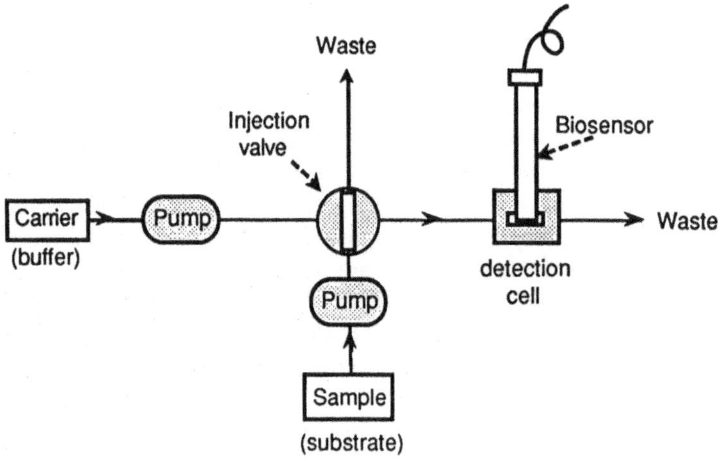

Figure 8.2 *Flow injection analysis using a biosensor for substrate determination.*

The FIA apparatus consists of two pumps regulated by a microcomputer (to transport the carrier liquid and inject the sample), an injection valve with a loop (which receives sample and carrier liquid alternately), a biosensor, a detection cell and linking tubes with an internal diameter of 0.5-1 mm. In this method, time, temperature and dispersion are all reproducible conditions. Unlike batch measurement, the response curve never reaches the plateau corresponding to the steady state but takes the form of peaks whose heights are related to the concentration of the analyte (Figure 8.3). The concentration in the sample is then determined by comparison with known standards which have been injected under the same conditions.

FIA has the advantage of working for low sample volumes (25-200 μl) and at a speed that is limited only by the response time of the biosensor. Biosensors with a rapid response [28] thus increase the sample throughput. If the speed at which the sample passes over the biosensor is considered to be constant, then a rapid biosensor response produces a larger peak height in FIA (Figure 8.3), which improves the signal to noise. With this aim in mind, an enzyme sensor with a response time of less than 10 seconds was coupled with FIA to measure penicillin [268]. Such a device avoids the need for an enzymatic reactor to convert penicillin into a product that is detectable by a transducer. The rapidity of the response of this biosensor can give a sample throughput as high as 200 measurements/hour.

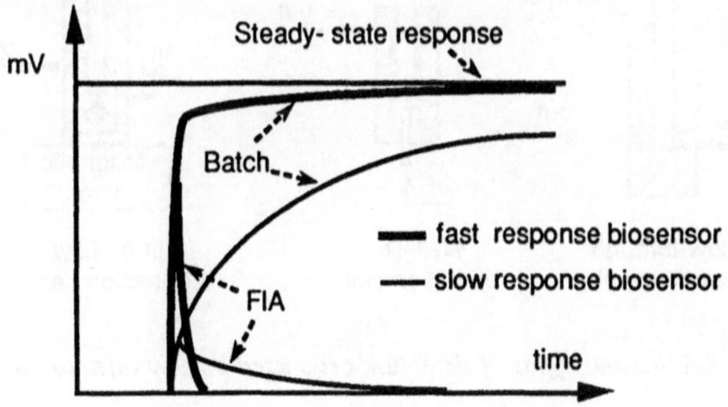

Figure 8.3 *Biosensor response in batch measurement and flow injection analysis.*

It is possible to use various types of detection cells (Figure 8.4), a flow-through detection cell, a wall-jet detection cell or a stirred-flow detection cell, each of which is designed for a specific application. The flow-through and wall-jet cells have very low dead volumes which permit rapid determination and a high sample throughput. For example, under the same conditions, penicillin was determined at a rate of 200 samples/hour with a flow-through cell, 175 samples/hour with a wall-jet cell, and only 40 samples/hour with a stirred-flow detection cell [268]. The wall-jet detection cell employs a jet of liquid aimed at the active membrane, which avoids the problems of deposition of suspended solids and fouling of the biosensor. However, this mechanical effect reduces the lifetime of certain immobilized biological compounds. The stirred-flow detection cell is better adapted for monitoring fermentation processes which do not require frequent measurement but which occur in a complex and changing medium where large variations in concentration may occur. The large dead volume of this cell facilitates the dilution of the broth, the pH can be maintained at a constant level, and the problems of interference reduced.

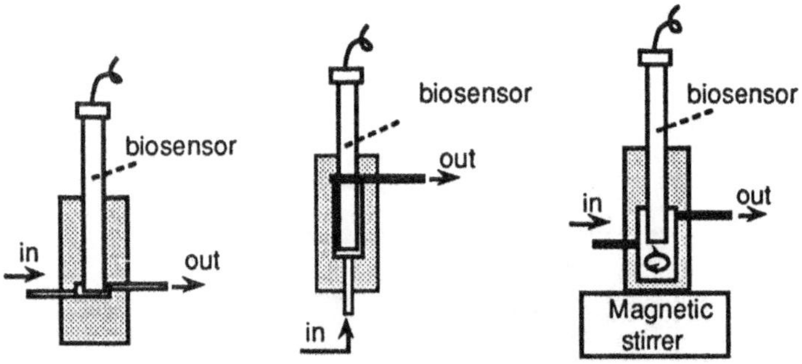

Flow-through detection cell

Wall-jet detection cell

Stirred-flow detection cell

Figure 8.4 *Various types of detection cells used in flow injection analysis.*

When a very small quantity of sample is injected into a large volume of liquid then dispersion must be taken into account. Dispersion is represented by:

$$D = \frac{C_0}{C_{max}}$$

where D is the dispersion coefficient, C_0 is the concentration of the product detected in the steady state (the batch method), and C_{max} is the concentration of the product represented by the height of the FIA peak obtained.

Figure 8.5 represents the variation of the dispersion coefficient as a function of the volume of the injected sample. The dispersion diminishes as the sample volume increases and the measurement becomes more sensitive but the sample throughput drops. A steady state is attained when the entire volume of the cell is occupied by the sample (D —> 1).

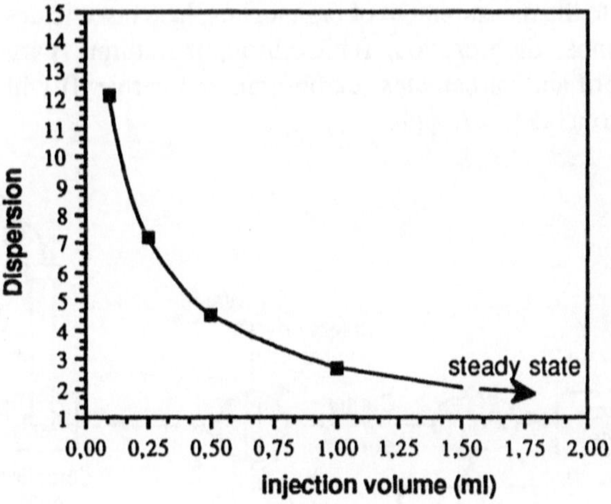

Figure 8.5 *Dispersion coefficient versus injected volume in flow injection analysis using a stirred-flow cell.*

(b) Determination of inhibitors

The determination of an inhibitor of an enzymatic reaction requires the presence of a substrate. There are two possibilities: the inhibitor could be injected into a carrier stream containing the substrate; or the substrate could be injected into a carrier stream containing the inhibitor. The first method only applies when the inhibition reaction is rapid (for example, the inhibition of urease by fluoride ions [132]) and has the disadvantage of a high consumption of substrate. The second method, injection of the substrate, is more widely used because inhibition reactions are often slow, especially those involving irreversible inhibitors. This method facilitates the incubation of the enzyme in the presence of its inhibitor and its reactivation by a regenerating agent (Figure 8.6). This system is particularly useful for monitoring water pollution because the toxic chemicals or polluting compounds that act as enzyme inhibitors are often found in flowing liquids. Furthermore, the low consumption of substrate make the method economically feasible and very convenient. The detection involves immobilization of the enzyme either on a membrane fixed to the electrode or in a reactor placed just upstream of the transducer [269]. This technique has been applied to the measurement of organophosphate insecticides (azinphos, bromophos, dichlorovos, fenitrothion, malathion, paraoxon, and parathion) and carbamates (carbofuran and carbaryl) with detection limits from 0.5 to 275 ppb.

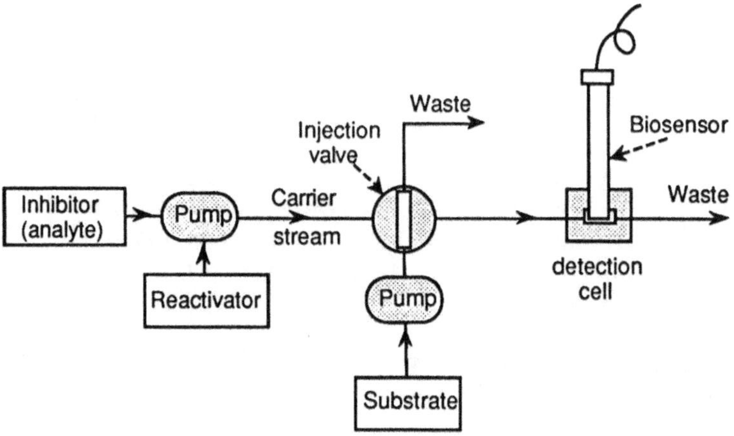

Figure 8.6. *Flow injection analysis using a biosensor for inhibitor determination.*

The system also presents the possibility of performing successive calibration, measurement and rinsing steps. The system is calibrated by first passing a carrier liquid with no inhibitor through the detection cell to obtain a reference peak corresponding to the maximal response of the biosensor. A series of solutions with different inhibitor concentrations are then used to obtain the corresponding percentage inhibitions (I %). Percentage inhibition is given by the following expression:

$$I\ \% = \frac{E_0 - E_I}{E_0} \times 100$$

where E_0 is the reference peak height given by the biosensor before inhibition and E_I is the peak height after inhibition (Figure 8.7).

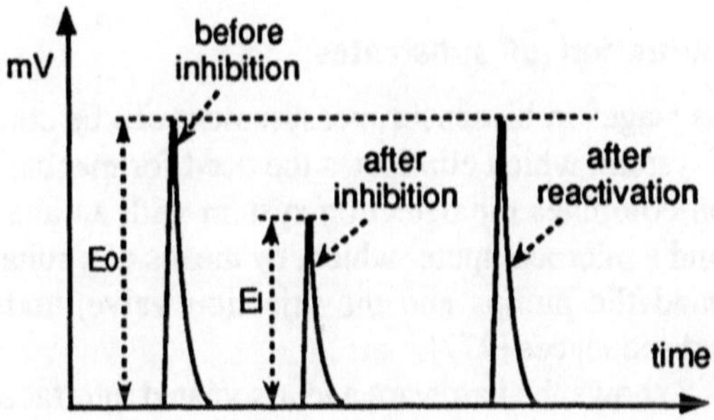

Figure 8.7. *Typical flow-injection peaks during inhibitor determination.*

Once the enzyme sensor has been in contact with the inhibitor, it is rinsed with a solution containing a reactivating agent. In principle, the injection of substrate should give the reference peak again.

When enzyme sensors are combined with FIA they have applications in medicine and the food produce industry. For example, FIA has been combined with the glucose oxidase electrode to determine glucose in blood [270], with the choline oxidase electrode for the determination of cholinesterase in serum [271], and with a bienzyme electrode that uses choline oxidase and phospholipase for the determination of phospholipids in serum [272]. The technique has also been used with an alcohol oxidase electrode to determine alcohol in drinks (for example, beers) [273], or with a glucose oxidase electrode to determine glucose [274] in various foodstuffs [275] or in fermentation processes [276].

The application of FIA and biosensors in environmental analysis is attractive because enzyme biosensors operate on the principle of inhibition. The process is economical because only small quantities of substrate are required and because the incubation, reaction and reactivation times can be controlled. The process can therefore yield high reproducibility. This method has been used to measure insecticide levels in sea water [269]. Finally, FIA is particularly suitable for the construction of automated systems.

8.1.3 Automation

(a) Determination of substrates

The various stages in biosensor measurement can be controlled by an automated system, which eliminates the need for manual intervention. Automation combines the detection system with an analog to digital converter and a microcomputer which, by means of a suitable interface, can command the pumps and the injection valve, and manage the programmed sequences [277].

Figure 8.8 shows the hardware and associated interfaces involved in an FIA measuring system that uses a penicillinase electrode to follow the production of penicillin by fermentation. This automated apparatus gives analogous results to those obtained by HPLC [277].

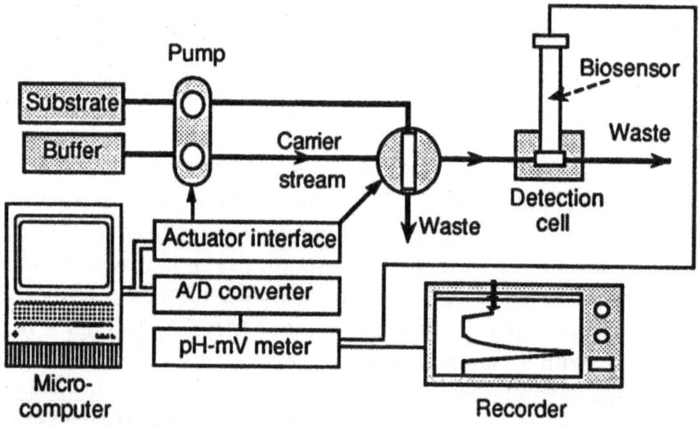

Figure 8.8 *Automation of flow injection analysis to monitor penicillin in fermentation broths.*

(b) Determination of inhibitors

One example of analysis using enzymatic inhibition is the determination of pesticides by the inhibition of a cholinesterase electrode. A simplified flow-chart representing the automation of this biosensor is given in Figure 8.9. The calibration and measuring steps are followed by the test for inhibition and then regeneration. This is characteristic of measurement by inhibition.

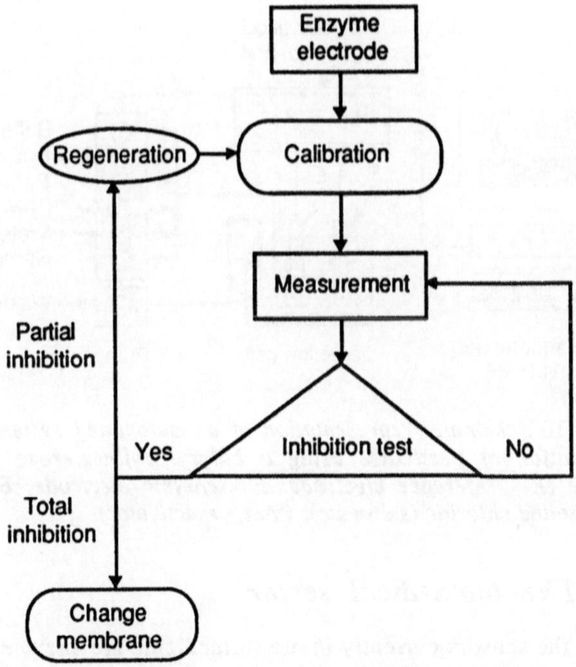

Figure 8.9 *Simplified flow-chart representing the operation of the automated determination of pesticides by an enzyme sensor.*

The total automated system is shown schematically in Figure 8.10. A number of pneumatic pumps send a series of known volumes of buffer solution, substrate, calibration inhibitor, regenerating agent, and rinsing solutions to the electrode [278]. The biosensor is calibrated, measures the concentration of the sample, is regenerated and washed, all in an automated fashion. This apparatus has been used to measure the toxicity of river water [279].

8.2 Applications

Biosensors have many commercial applications in a large range of activities. The most important applications are in medicine (in hospitals or in the home) and in the food produce industry for the control of manufacturing processes. Other applications include the protection of the environment and defense.

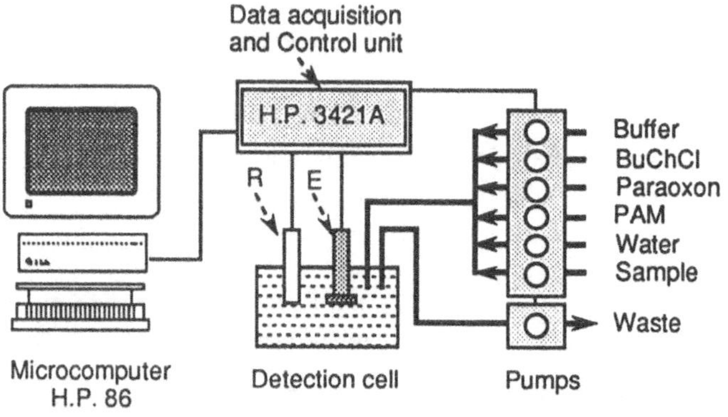

Figure 8.10 *Schematic representation of an automated system for the determination of pesticides using a butyrylcholinesterase (BuChE) electrode (R - reference electrode; E - enzyme electrode; BuChCl - butyrylcholine chloride (substrate); PAM - reactivator).*

8.2.1 The biomedical sector

Most of the sensors currently in use in medicine are *enzyme sensors* because of the specificity of their measurement, their ease of implementation, and the fact that the enzymes and their corresponding transducers are commercially available. The glucose oxidase sensor has received the most attention due to its potential for the determination of glucose in blood and urine for the diagnosis of diabetes. This is also the reason why so much research has been devoted to this sensor (Table 4.3). At present, glucose levels in the blood of diabetics are measured by one of two systems. One of these uses an open-loop device, through which insulin is injected into the patient following a predetermined programme, with no allowance for glucose concentration. The second system uses a closed-loop device through which insulin, or other substances, are injected as a function of the glucose concentration at any one time. The enzyme electrode monitors glucose levels in the blood continuously, and the system can be regarded as a sort of artificial pancreas [280]. The in-vivo use of biosensors requires the resolution of the problems of biocompatibility, especially those concerning the deposit of fibrin and platelets at the enzymatic membrane [281].

Biosensors other than the glucose electrode are capable of measuring metabolites in biological media. Urea and creatinin electrodes can control renal function; the cholesterol electrode can be used for the detection and prevention of arteriosclerosis; the acetylcholine electrode can monitor the neurotransmitters related to chemical transmission at the synapses; and the lactate electrode can evaluate muscular effort [282].

The applications of *microbial sensors* in medicine are limited because they are unsuitable for use in biological liquids, which are usually the culture medium for microorganisms. The growth in biomass deforms the biocatalytic matrix leading to a leak of cells and a contamination of the sample medium. Furthermore, microbial sensors contain a large number of enzymes and are not sufficiently specific for many biomedical analyses.

Immunological sensors have a greater potential in medicine because of the specificity of immunological reactions. These sensors are used to measure drugs such as theophylline, and to determine the hormone HCG for the diagnosis of pregnancy, alpha-fetoprotein for the identification of cancer, and the surface antigen of hepatitis B. However, these sensors cannot be used *in vivo* because the enzymatic amplification involved requires the addition of a substrate for the operation of the sensor. Moreover, the formation of the antigen-antibody complex is slow and requires a number of steps. The immunological sensor works in sequence and cannot provide continuous control.

8.2.2 The food produce industry

Procedures in the food produce industry are becoming more and more automated. There is a large demand for biosensors to monitor the different stages in production and the quality control of the final product.

The sensitivity of biosensors towards biological products means that they are particularly suitable for monitoring ingredients, food additives, contaminants, and toxins. A glucose electrode can determine the freshness of meat by the glucose consumption of the microbes on its surface [283]. Similarly, the freshness of fish is indicated by a sensor that is sensitive to hypoxanthine, inosine and inosine-5'-monophosphate [284, 285]. There are multiple uses for biosensors in the food produce industry, such as the determination of lactate for the quality control of wine and yoghurt [286], and of glucose and penicillin for monitoring fermentation [166, 268, 287, 288]. Further examples include the determination of amino acids (such as lysine) for the control of proteins and the nutritional value of foodstuffs [121], of sucrose for the production of syrups and jams [4], and of ethanol for the evaluation of the degree of alcohol in alcoholic drinks [170]. Enzyme sensors are best suited to these requirements, as was the case for medical applications.

The use of biosensors in fermentation processes still poses a number of problems. Biosensors are based on enzymatic reactions and constitute an invasive method since they liberate small quantities of reaction products. In order to eliminate this, the biosensor is placed in a derivative loop and the portions of sample that are measured are not recycled. This apparatus also avoids problems of contamination in the fermenter. However, the extraction does mean a loss of culture medium and the sample collected is not always representative of the entire fermentation broth.

Other trials have involved placing the biosensor, without its biocatalyst, in an autoclave [287]. The enzyme is then injected into a microchamber where it remains during the operation of the biosensor. The enzyme can be renewed when its activity is judged to be insufficient.

8.2.3 Environmental protection

The quality of the environment in urban areas is related to the extent of the control of atmospheric and water pollution. Water pollution arises from both domestic and industrial waste, and biosensors can play a valuable role here, in measuring organic pollution or evaluating toxic substances.

(a) Measurement of organic pollution

Organic pollution is measured by a well-established and standardized test which records the biological oxygen demand (BOD). The test requires sample incubation over 5 days at 20 °C. The BOD measurement can be made in a much shorter time (about 15 minutes) with a microbial biosensor [62]. This sensor is based on the immobilization of whole hydrogen-producing bacteria cells (*Clostridium butyricum*) on the oxygen-permeable membrane of a pO_2

electrode. Once the sample is saturated with oxygen, it is placed in the presence of the organic compound and the consumption of oxygen by the bacteria causes a drop in the electrode signal. The value of the BOD is then determined with the help of a calibration curve.

(b) Measurement of toxicity

Toxicity is difficult to evaluate owing to the plurality of substances that are more or less toxic. Total toxicity is usually measured using aquatic animals that are very sensitive to pollution, for example, trout. It is, however, difficult to detect the death of one or more trout automatically, without interference from the risk of false alarms such as accidental death [289]. There are other biological tests, such as the test for daphnia mobility and Microtox bacteria tests, which use large populations of living organisms to improve reliability. These biological tests remain difficult to automate, hence the research into biosensors that are sensitive to total toxicity.

Toxic compounds act through enzymatic inhibition and the immobilization of target enzymes on transducers provides a continuous measurement of the activity of the biocatalyst. The biosensor reproduces the poisoning of living organisms, and detects the actual toxicity level [290].

The cholinesterase electrode was constructed on this basis for the total determination of organophosphates and carbamates [32]. This electrode is also sensitive to other toxic compounds with very low detection limits (paraquat (3 ppm); trichlorophenol (20 ppm); methylazinphos (150 ppb); lindane (15 ppm); and mercury (4 ppm)) [135]. This type of biosensor uses enzymatic inhibition and can be incorporated in an automated system for the quality control of water, and, notably, waste water [291].

8.2.4 Defense

Military leaders have long been preoccupied with improving the protection of their armed forces against nuclear and chemical warfare. Chemical agents are extremely flexible and thus make formidable weapons. On the whole, the most toxic agents are those derived from organophosphate pesticides, which can be classed into two categories, the "G" agents (GA (tabun), GB (sarin) and GD (soman)), which act in the respiratory tract, and the "V" agents (VG and VX), which penetrate the skin. These agents inhibit acetylcholinesterase, one of the key enzymes of the nervous system.

The varied nature of neurotoxic compounds makes them difficult to detect by physicochemical methods, and so it is important to find a biosensor that is capable of detecting them. Cholinesterase is one of the target enzymes of these derivatives and can be used to construct a biosensor that is sensitive to extremely low quantities of neurotoxic agents [292]. The biosensor includes a potentiometric transducer and a membrane containing the immobilized enzyme. In the absence of any poisons, a reference signal is obtained which corresponds to the maximal activity of the enzyme. When the sensor is in contact with the neurotoxic agent, a complex is formed between the organophosphate and the active site of the enzyme, the serine group. This competitive inhibition blocks the function of the enzyme [56]. The lowering of the enzymatic activity leads to a drop in the biosensor signal which is proportional to the concentration of the neurotoxic agent. The sensor can be connected to an alarm system to give good protection against chemical warfare agents.

Cholinesterase is an important target enzyme and research continues on its coupling with other transducers. One method is to measure the variation in frequency of a quartz crystal with cholinesterase immobilized on its surface [228]. Neurotoxic agents are detected by the variation in mass caused by their fixation to the enzymatic quartz. An antibody specific to a particular organophosphate can be immobilized on the quartz crystal in the place of the enzyme. The organophosphate will then interact with the immunological quartz. This system has the disadvantage of being too specific, the biosensor is incapable of detecting organophosphates with similar structures, and the effectiveness of its protection is limited. Other transducers currently under investigation include optical fibers which have advantages for use in hostile environments (electric or magnetic fields, or explosive atmospheres).

9

Economic factors

The various applications described in Chapter 8 indicate the potential market for biosensors. The resulting economic factors are important and should be considered. The pH electrode is now commonly used both in the laboratory and on site, and the biosensor appears to be following a similar progression. The glucose electrode had already made a significant impact on the biomedical sector only a short time after it was placed on the market. In three years, 800 Yellow Springs Instrument Company analysers (model 23 A) were sold to different medical analysis laboratories [293]. This success arises from the fact that the price per measurement dropped considerably. Analysers in which the enzyme was used once and then discarded have been replaced by analysers in which a small quantity of enzyme is immobilized on a membrane, but which allow thousands of measurements per membrane. Owens-Illinois (Toledo) transferred the patent rights to the company Technicon, in order to exploit the coupling of immobilized glucose oxidase with a platinum electrode for the measurement of glucose [294].

Japan dominates the present worldwide effort in biosensor research and development (R&D). The Japanese market in biosensor products is currently estimated at 1.5 million yen. About a hundred different firms from a variety of fields consider biosensors to be an important research direction (for example, Toyo Jozo, Fuji Electric, Toyobo, and Denki) or form collaborations for better development (for example, Daikin of Sumitomo and NEC). In Germany, 15 million D.marks were placed for financing biosensor research and development between 1988 and 1991. Other European countries (Great Britain, France, Italy and Spain) also participate actively in research in this field.

The originality of biosensors is their multifaceted character. Consider the glucose electrode, which is used in medical analysis for in-vitro measurements, and implanted in diabetics to regulate insulin. It is also employed to control meat freshness, and sugar levels in yoghurt, wine, fruit juices, and other foodstuffs, and to control and regulate fermentation where glucose is often the substrate for the microorganisms. Finally, the same electrode can be used in environmental protection by monitoring the waste water from the food produce industry (hydrolysis of lactate in cheese production).

Of course the glucose electrode is just one example of this multifaceted character. Biosensors are used extensively in process control in industry and permit automation. They eliminate sampling and analysis steps, thereby reducing the amount of human intervention and the number of personnel, and increasing the procedure reliability and the product quality. On the whole, the use of biosensors is attractive because they lead to a higher quality product at a more competitive price.

The actual implementation of a biosensor does not pose any major problems. The components, the bioreceptor and the transducer, are commercially available, or can be easily obtained. Furthermore, electrochemical transducers (pH, pO_2, pCO_2 and pNH_3 electrodes) use cheap and reliable amplification devices. A pH electrode together with a portable measuring system presently costs about $200. Biosensors are moving in the same direction, with an increase in the reliability and a reduction in cost. The consequences of this are a more perceptible penetration of biosensors into the different sectors concerned and an interest in new applications.

There is presently a move towards biosensors that are multiparametric, miniaturized, and portable. The pharmaceutical industry is particularly difficult to penetrate; there are a limited number of bioreactors, the products are expensive, and the analytical methods used are often very sophisticated (for example, chromatography or HPLC). In contrast, biosensors have a large number of applications and a promising future in the areas of environment and defense, in which a number of sensors in different sites can increase the efficiency of protection.

Small biotechnology companies can use biosensors to control the quality of their products or to automate certain steps in their processes. A medical analysis laboratory can replace some existing techniques by biosensors, provided that this reduces the cost, and hence the investment. Biosensors can also be used to increase the value of products. For example, manufacturers of fermenters can supplement their products with biosensors (glucose or penicillin), and manufacturers of analysers or automated analysis systems can use one or more biosensors thereby increasing the performance of their products.

At present most biosensors are produced by the manufacturers of transducers because making the biosensor simply involves finding the appropriate bioreceptor and its operating system. The immobilization of the bioreceptor on the transducer is not a major problem, especially taking into account the diversity of the methods available. A manufacturer of pH, pO_2, pCO_2 and pNH_3 electrodes is capable of producing a large number of biosensors based on electrochemical detection. It would, however, be difficult for such a manufacturer to produce ENFET transducers without access to semiconductor technology. The lack of the appropriate transducer technology is usually the factor that prevents wider development of biosensors. Problems in amplification and signal processing are rare because these are common to all categories of transducers.

Another obstacle is the low sale price of biosensors (presently about $300-$500) which does not encourage large investment for their study and perfection. Manufacturers usually sell biosensors included in analysers, or integrated into an automated system for analysis. The attraction of biosensors is thus to diversify the products that the manufacturer can offer for sale within a well-established commercial network.

Although biosensors themselves are not very expensive, their research costs are often quite high. The field is multidisciplinary and a number of different scientists play major roles: biochemists, biologists, microbiologists, immunologists, medical scientists, chemists, physicists, electronic engineers, and computer scientists. The research also requires specific techniques that a small enterprise may not possess. There are a number of important points which may limit the use of biosensors on a large scale, notably their lifetime and stability. Research effort should be orientated towards improvement in reliability, and the resolution of the problems of sterilization and biocompatibility of biosensors.

Training and technical assistance should be provided by both public and private laboratories of the universities and technical centers in which biosensors are under investigation. The laboratory recognizes the product, and investigates its operation, its possibilities, and its limits. A biosensor generally works in a specific pH range and close to ambient temperatures. There may be interference from other substances or problems with reversibility or loss of activity; the biosensor must fulfil all the conditions posed by its application. For example, a lifetime of 2-3 weeks, for use in bioreactors, and several months for monitoring diabetes in a patient.

In both the long and the short term, the benefits depend on the sale price of the apparatus. The benefits are certainly more significant when the biosensor is sold with its analyser, which is why each manufacturer often has a different method of linking the biosensor to the analyser. However, the price per measurement must be competitive, and must not exceed about a dollar a measurement in biomedical automated systems. The market for biosensors is vast, but it is vital to have an early presence on this market in which the competition is so stiff.

10

Conclusions

Biosensors have undergone rapid development over the last few years. This is due to the combination of new bioreceptors with the ever-growing number of transducers. The characteristics of these biosensors have been improved, and their increased reliability has yielded new applications.

The predominant application of biosensors is in the biomedical field which has a constant need for monitoring biological parameters in health care techniques. The biosensor is particularly suitable for such measurements because a sample can be analysed in a complex biological environment without the need for chemical reagents. Biosensors can be used *in vivo* and, because they give a continuous signal, can monitor metabolite concentrations in real time. Hence their important application in the control of blood sugar in diabetics.

The applications of biosensors in the food produce industry have developed in parallel. One major breakthrough is the possibility of sterilization, allowing the use of biosensors in fermentation processes. Biosensors also play an important role in direct control of foodstuffs.

The environment also requires continuous surveillance if it is to have proper protection but present physicochemical techniques are limited in this respect, especially with regard to toxicity. The same applies in defense. Biosensors can meet these needs, the target enzymes of the toxic agents are simply associated with the appropriate transducer.

Current research concentrates on improving biosensor sensitivity (through the use of mediators and enzymatic amplification) and selectivity (through the use of immunoagents). Their use comes up against many unresolved technical problems and a number of factors remain to be investigated, for example, miniaturization for in-vivo applications, biocompatibility, stability, and response time. The next steps will be decisive and the competition is very stiff. A large number of industrial companies in Japan are already involved in the development and manufacture of biosensors [295], whereas activity in Europe is still very limited. Action should be taken to make companies aware of this profitable field, where economic factors play such an important role.

References

[1] Clark L.C. and Lyons C. (1962) Electrode systems for continuous monitoring in cardiovascular surgery. *Ann. N.Y. Acad. Sci.*, **102**, 29-45.

[2] Updike S. J. and Hicks G.P. (1967) The enzyme electrode. *Nature* , **214**, 986-988.

[3] Tran-Minh C. and Broun G. (1975) Construction and study of electrodes using cross-linked enzymes. *Anal. Chem.*, **47**, 1359-1364.

[4] Satoh I., Karube I. and Suzuki S. (1976) Enzyme electrode for sucrose. *Biotech. Bioeng.*, **18**, 269-271.

[5] Hartmeier W. (1986) *Immobilized biocatalysts*. Springer Verlag, p. 49.

[6] Pandey P.C. (1988) A new conducting polymer coated glucose sensor. *J. Chem. Soc. Faraday Trans. I.*, **84**, 2259-2265.

[7] Caras S. and Janata J. (1980) Field effect transistor sensitive to penicillin. *Anal. Chem.*, **52**, 1935-1937.

[8] Goldfinch M.J. and Lowe C.R. (1984) Solid-phase optoelectronic sensors for biochemical analysis. *Anal. Biochem*, **138**, 430-436.

[9] Arnold M.A. (1985) Enzyme-based fiber optic sensor. *Anal. Chem.*, **57**, 565-566.

[10] Munkholm C., Walt D.R., Milanovich F.P. and Klainer S.M. (1986) Polymer modification of fiber optic-chemical sensors as a method of enhancing fluorescence signal for pH measurement. *Anal. Chem.*, **58**, 1427-1430.

[11] Freeman T.M. and Seitz W.R. (1978) Chemiluminescence fiber optic probe for hydrogen peroxide based on the luminol reaction. *Anal. Chem.*, **50**, 1242-1246.

[12] Tran-Minh C. and Vallin D. (1978) Enzyme bound thermistor as an enthalpimetric sensor. *Anal. Chem.*, **50**, 1874-1878.

[13] Ader J.F. and McCallum J.J. (1983) Piezoelectric crystals for mass and chemical measurements. *Analyst*, **108**, 1169-1189.

[14] Divies C. (1975) Remarques sur l'oxydation de l'éthanol par une électrode microbienne *d'Acetobacter xylinum. Ann. Microbiol. Inst. Pasteur*, **126A**, 175-186.

[15] Yamamoto N., Nagasawa Y., Shuto S., Sawai M., Sudo T. and Tsubomura H. (1978) The electrical method of investigation of the antigen-antibody and enzyme-enzyme inhibitor reactions using chemically modified electrodes. *Chem. Letters*, 245-246.

[16] Thompson M., Dorn W.H., Krull U.J., Tauskela J.S., Vandenberg E.T. and Wong H.E. (1986) The primary events in chemical sensory perceptor. *Anal. Chim. Acta*, **180**, 251-269.

[17] Arnold M.A. and Rechnitz G.A. (1987) Biosensors based on plant and animal tissue, in *Biosensors. Fundamentals and applications*. (ed. A.P.F. Turner, I. Karube and G.S. Wilson), Oxford Univ. Press, 30-59.

[18] Schubert F. and Scheller F.W. (1988) Organelle electrodes, in *Methods in Enzymology*, vol. 137. (Immobilized enzymes cells, Pt D), 152-160.

[19] Yamamoto N., Nagasawa Y., Sawai M., Sudo T. and Tsubomura H. (1978) Potentiometric investigations of antigen-antibody and enzyme-enzyme inhibitor reactions using chemically modified metal electrodes. *J. Immunol. Methods* , **22**, 309-317.

[20] Aizawa M., Morioka A., Matsuoka H., Suzuki S., Nagumura Y., Shinohara R. and Ishiguro I. (1976) An enzyme immunosensor for IgG, *J. Solid Phase Biochem.*, **1**, 319-328.

[21] Keating M.Y. and Rechnitz G.A. (1984) Potentiometric digoxin antibody measurements with antigen-ionophore based membrane electrodes. *Anal. Chem.*, **56**, 801-806.

[22] Wingard L.B. Jr. (1987) Possibilities for biosensors based on neuroreceptors, in *Biosensors international workshop*. (ed. R.D. Schmid *et al.*) GBF Monographs, vol. 10, 133-137.

[23] Guilbault G.G. and Shu F.R. (1972) Enzyme electrodes based on the use of a carbon dioxide sensor. Urea and L-Tyrosine electrodes. *Anal. Chem.*, **44**, 2161-2165.

[24] Mascini M. and Guibault G.G. (1986) Clinical uses of enzyme electrode probes. *Biosensors*, **2**, 147-172.

[25] Suaud-Chagny M.F. and Pujol J.F. (1985) Microélectrode à enzyme immobilisée pour la détection de l'acétylcholine. *Analusis*, **13**, 25-29.

[26] Beaux J. and Tran-Minh C. (1979) Inhibition de l'acétylcholinestérase immobilisée par les ions fluorure. *C. R. Acad. Sci. Paris, Ser. C*, **288**, 545-548.

[27] Kobos R.K., Eveleigh J.W., Stepler M.L., Haley B.J. and Papa S.L. (1988) Fluorocarbon based immobilization method for preparation of enzyme electrodes. *Anal. Chem.*, **60**, 1996-1998.

[28] Kumaran S., Meier H., Danna A.M. and Tran-Minh C. (1991) Immobilization of thin membranes for the construction of glass enzyme electrodes. *Anal. Chem.*, **63**, 1914-1918.

[29] Tran-Minh C., Kumaran S., Meier H. and Danna A.M. (1990) Immobilisation des protéines. F. *patent*. 901 1630.

[30] Tran-Minh C. and Broun G. (1973) Utilisation des électrodes à pression partielle de gaz carbonique pour la mesure spécifique de l'urée en solution. *C.R. Acad. Sci. Paris, Ser. D*, **276**, 2215-2217.

[31] Mascini M., Iannillo M. and Palleschi G. (1983) Enzyme electrodes with improved mechanical and analytical characteristics obtained by binding enzymes to nylon nets. *Anal. Chim. Acta*, **146**, 135-148.

[32] Tran-Minh C. and El Yamani H. (1988) Enzyme sensors for determination of toxic chemicals in environmental samples, in *Electrochemical detection techniques in the applied biosciences* (ed. G.A. Junter), vol. 2, Ellis Horwood, 131-141.

[33] Tran-Minh C. and Guyonnet R. (1987) Capteurs à enzymes. F. *Patent* 8707589.

[34] Tran-Minh C. and Kumaran S. (1990) Comparative study of different enzyme immobilisation techniques suitable for inhibitor determination. *Biosensors '90*, Singapore.

[35] Miyabayashi A. and Mattiasson B. (1988) An enzyme electrode based on electromagnetic entrapment of the biocatalyst bound to magnetic beads. *Anal. Chim. Acta*, **213**, 121-130.

[36] Watanabe E., Endo H. and Toyama K. (1988) Determination of phosphate ions with an enzyme sensor system. *Biosensors*, **3**, 297-306.

[37] Scheller F.W., Schubert F., Weigelt D., Mohr P. and Wollenberger U. (1988) Molecular recognition and signal processing in biosensors. *Makromol. Chem., Macromol. Symp.*, **17**, 429-439.

[38] Thomas J.D.R. (1988) Uni-, bi- and tri-enzyme electrodes for analysis. *NATO ASI Ser., Ser. C*, **226**, 141-152.

[39] Pilloton R., Mascini M., Casella I.G., Festa M.R. and Bottari E. (1987) Lactose determination in raw milk with two enzyme based electrochemical sensor. *Anal. Letters*, **20**, 1803-1814.

[40] Yoda K. (1988) Multienzyme membrane electrodes, in *Methods in Enzymology*, vol. 137. (Immobilized enzymes cells, Pt D), 61-68.

[41] Mosbach K. (1978) Immobilized coenzymes in general ligand affinity chromatography and their use as active coenzymes. *Adv. Enzymology*, **46**, 205-278.

[42] Ukeda H., Kamikado H., Matsumoto K. and Osajima Y. (1989) A new approach of the coimmobilization of alcohol dehydrogenase and NAD on glutaraldehyde activated Sepharose and its application to the enzymatic analysis of ethanol. *Agric. Biol. Chem.*, **53**, 25-30.

[43] Bartlett P.N. and Whitaker R.G. (1988) Strategies for the development of amperometric enzyme electrodes. *Biosensors*, **3**, 359-379.

[44] Tran-Minh C., Pandey P.C. and Lantreibecq F (1991) Electrochemical studies on tetrathiafulvalene tetracyanoquinodimethane modified acetylcholine/choline sensor. *Appl. Biochem. Biotech.*, **31**, 145-158.

[45] Foulds N.C. and Lowe C.R. (1988) Immobilization of glucose oxidase in ferrocene-modified polypyrrole polymers. *Anal. Chem.*, *60*, 2473-2478.

[46] Fortier G., Brassard E. and Bélanger D. (1988) Fast and easy preparation of an amperometric glucose biosensor. *Biotechnol. techniques*, **2**, 177-182.

[47] Fortier G., Chen J.W. and Bélanger D. (1992) Biosensors based on entrapment of enzymes in a water dispersed anionic polymer, in *Biosensors and chemical sensors, ACS Symposium Ser.* **487** (**2**), 22-30.

[48] Hale P.D., Inagaki T., Karan H.I., Okamoto Y. and Skotheim T.A. (1989) A new class of amperometric biosensor incorporating a polymeric electron-transfer mediator. *J. Am. Chem. Soc.*, **111**, 3482-3484.

[49] Tran-Minh C., Kumaran S., Abdul M. and El Yamani H. (1990) Pd/PdO based urea and butyrylcholine microbiosensors. *Anabiotec '90*, 3rd Int. Symposium on analytical methods and problems in biotechnology, Burlingame, San Francisco.

[50] Tran-Minh C., Pandey P.C. and Kumaran S. (1990) Studies on acetylcholine sensor and its analytical application based on the inhibition of cholinesterase. *Biosensors Bioelectronics*, **5**, 461-471.

[51] Karube I. and Moriizumi T. (1988) Microenzyme sensors, in *Methods in Enzymology*, vol. 137. (Immobilized enzymes cells, Pt D), 255-260.

[52] Karube I. and Kubo I. (1988) Microenzyme biosensors. *NATO ASI Ser., Ser. C*, **226**, 207-218.

[53] Shiono S., Hanazoto Y., Nakako M. and Maeda M. (1987) Photolithographic patterning of enzyme immobilized membrane and new flow-through cell for enzyme-modified FET biosensor. *Biosensors international workshop*. (ed. R.D. Schmid *et al.*) GBF Monographs, vol. 10, 291-292.

[54] Bourdillon C., Bourgeois J.P. and Thomas D. (1980) Covalent linkage of glucose oxidase on modified glassy carbon electrodes. Kinetic Phenomena. *J. Am. Chem. Soc.*, **102**, 4231-4235.

[55] Shichiri M., Kawamori R. and Yamasaki Y. (1987) Needle-type glucose sensor and its clinical applications. *Biosensors. Fundamentals and applications*. (ed. A.P.F. Turner, I. Karube and G.S. Wilson), Oxford Univ. Press, 409-424.

[56] Tran-Minh C. (1985) Immobilized enzymes probes for determining inhibitors. *Ion-Selective Electrode Reviews*, **7**, 41-75.

[57] Suaud-Chagny M.F. and Gonon F.G. (1986) Immobilization of lactate dehydrogenase on a pyrolytic carbon fiber microelectrode. *Anal. Chem.*, **58**, 412-414.

[58] Ikariyama Y., Yamauchi S., Yukiashi T. and Ushioda H. (1989) Electrochemical fabrication of amperometric microenzyme sensor. *J. Electrochem. Soc.*, **136**, 702-706.

[59] Matsunaga T., Karube I. and Suzuki S. (1978) Rapid determination of nicotinic acid by immobilized *Lactobacillus arabinosus*. *Anal. Chim. Acta*, **99**, 233-239.

[60] Rechnitz G.A., Kobos R.K., Riechel S.J. and Gebauer C.R. (1977) A bioselective membrane electrode prepared with living bacterial cells. *Anal. Chim. Acta*, **94**, 357-365.

[61] Walters R.R., Moriarty B.E. and Buck R.P. (1980) *Pseudomonas* bacterial electrode for determination of L-histidine. *Anal. Chem.*, **52**, 1680-1684.

[62] Karube I., Matsunaga T., Mitsuda S. and Suzuki S. (1977) Microbial electrode BOD sensors. *Biotech. Bioeng.*, **19**, 1535-1547.

[63] Karube I., Matsunaga T. and Suzuki S. (1979) Microbioassay of nystatin with a yeast electrode. *Anal. Chim. Acta*, **109**, 39-44.

[64] Kobos R.K., Rice D.J. and Flournoy D.S. (1979) Bacterial membrane electrode for the determination of nitrate. *Anal. Chem.*, **51**, 1123-1125.

[65] Riechel T.L. and Rechnitz G.A. (1978) Hybrid bacterial and enzyme membrane electrode with NAD response. *J. Membrane Sci.*, **4**, 243-250.

[66] Hikuma M., Kubo T., Yasuda T., Karube I. and Suzuki S. (1980) Ammonia electrode with immobilized nitrifying bacteria. *Anal. Chem.*, **52**, 1020-1024.

[67] Suzuki H., Tamiya E. and Karube I. (1989) A hybrid L-tyrosine sensor using an enzyme and bacterial CO_2 sensor. *Anal. Letters*, **22**, 15-24.

[68] Mascini M., Memoli A. and Olana F. (1987) Electrochemical biosensors for determination of nystatin activity. *Anal. Chim. Acta*, **200**, 237-244.

[69] Karube I., Tamiya E., Jode K., Yokoyama K., Kitagawa Y., Suzuki H. and Asano Y. (1988) Application of microbiological sensor in fermentation processes. *Anal. Chim. Acta*, **213**, 69-77.

[70] Mascini M., Memoli A. and Olana F. (1989) Microbial sensor for alcohol. *Enzyme Microb. Technol.*, **11**, 297-301.

[71] Robinson G.A., Cole V.M., Rattle S.J. and Forest G.C. (1986) Bioelectrochemical immunoassay for human chorionic gonadotropin in serum using an electrode-immobilised capture antibody. *Biosensors*, **1**, 45-57.

[72] Rosen I. and Rishpon J. (1989) Alkaline phosphatase as a label for a heterogeneous immunoelectrochemical sensor. An electrochemical study. *J. Electroanal. Chem.*, **258**, 27-39.

[73] Kooyman R.P.H., Kolkman H., Van Gent J. and Greve J. (1988) Surface plasmon resonance immunosensors: sensitivity considerations. *Anal. Chim. Acta*, **213**, 35-45.

[74] Aizawa M., Morioka A., Suzuki S. and Nagamura Y. (1979) Amperometric determination of human chorionic gonadotropin by membrane bound antibody. *Anal. Biochem.*, **94**, 22-28.

[75] Aizawa M., Morioka A. and Suzuki S. (1978) Electrochemical determination of IgG with an antibody bound membrane. *J. Membrane Sci.*, **4**, 221-228.

[76] Aizawa M., Morioka A. and Suzuki S. (1980) An enzyme immunosensor for the electrochemical determination of the tumor antigen α-fetoprotein. *Anal. Chim. Acta*, **115**, 61-67.

[77] Mattiasson B. and Nilsson H. (1977) An enzyme immunoelectrode. Assay of human serum albumin and insulin. *FEBS Letters*, **78**, 251-254.

[78] Bush D.L. and Rechnitz G.A. (1988) Comparison of antibodies as molecular recognition elements for biosensor design. *Anal. Letters*, **21**, 1947-1967.

[79] Karube I. (1988) Immunosensors. *NATO ASI Ser., Ser. C*, **226**, 267-279.

[80] Tromberg J.B., Sepaniak M.J., Alarie J.P., Vo-Dinh T. and Santella R.M. (1988) Development of antibody-based fiber-optic sensors for detection of a benzo [a] pyrene metabolite. *Anal. Chem.*, **60**, 1901-1908.

[81] Solsky R.L. and Rechnitz G. A. (1981) Preparation and properties of an antibody selective membrane electrode. *Anal. Chim. Acta*, **123**, 135-141.

[82] Keating M.Y. and Rechnitz G.A. (1983) Cortisol antibody electrode. *Analyst*, 766-768.

[83] Tran-Minh C., Pandey P.C. and Chavanne D. (1992) A new membrane electrode for the detection of antibody. *Biosensors Bioelectronics*, **7**, 147-149.

[84] Ikariyama Y., Furuki M. and Aizawa M. (1982) Bioaffinity sensor with binding protein. *Anal. Chem. Symp. Series*, **17**, 693-698.

[85] Ikariyama Y. and Aizawa M. (1988) Bioaffinity sensors, in *Methods in Enzymology*, vol. 137. (Immobilized enzymes cells, Pt D), 111-124.

[86] Meyerhoff M.E. and Rechnitz G. A. (1979) Electrode based enzyme immunoassays using urease conjugates. *Anal. Biochem.*, **95**, 483-493.

[87] Schubert F., Wollenberger U. and Scheller F. (1983) Plant tissue-based amperometric tyrosine electrode. *Biotechnol. Letters*, **5**, 239-242.

[88] Rechnitz G. A. (1988) Bioselective membrane electrodes using tissue materials as biocatalysts, in *Methods in Enzymology*, vol. 137. (Immobilized enzymes cells, Pt D), 138-152.

[89] Uchiyama S., Tamata M., Tofuku Y. and Suzuki S. (1988) A catechol electrode based on spinach leaves. *Anal. Chim. Acta*, **208**, 287-290.

[90] Wang J. and Lin M.S. (1988) Mixed plant tissue-carbon paste bioelectrode. *Anal. Chem.*, **60**, 1545-1548.

[91] Belli S.L. and Rechnitz G.A. (1988) Biosensors based on native chemoreceptors. *Fresenius Z. Anal. Chem.*, **331**, 439-447.

[92] Buch R.M. and Rechnitz G.A. (1989) Neuronal biosensors. *Anal. Chem.*, **61**, 533A-542A.

[93] Valdes J.J., Wall J.G.Jr., Chambers J.P. and Eldefrawi M.E. (1988) A receptor-based capacitive biosensor. *Johns Hopkins APL Technical Digest*, **9**, 4-10.

[94] Carr W.P. and Bowers L.D. (1980) Immobilized enzymes in analytical and clinical chemistry. *Chemical Analysis.*, John Wiley and sons, vol. 56, chap. V, 197-310.

[95] Beaux J. (1983) *Etude du comportement des enzymes par capteurs enzymatiques. Activités catalytiques et phénomènes d'inhibition. Analyse mathématique et applications analytiques.* PhD Thesis, Grenoble, p. VII.9.

[96] Goldman R., Goldstein L. and Katchalski E. (1971) *Biochemical aspects of reactions on solid supports.* (ed. G.R. Stark) Academic Press, New York, p. 1.

[97] Barman T.E. (1969) *Enzyme handbook*, Springer Verlag, Berlin, vol. II, p. 649.

[98] Guilbault G.G. and Montalvo J.G. Jr. (1969) A specific urea electrode. *J. Am. Chem. Soc.*, **91**, 2164-2165.

[99] Tran-Minh C., Sélégny E. and Broun G. (1972) Réalisation et étude du comportement d'une électrode à uréase réticulée. *C.R. Acad. Sci. Paris, Ser. C*, **275**, 309-312.

[100] Commission on analytical nomenclature (1976) Recommendations for nomenclature of ion-selective electrodes. *Pure and Appl. Chem.*, **48**, 127-132.

[101] Tran-Minh C., Guyonnet R. and Beaux J. (1978) Réalisation et étude d'une électrode à acétylcholinestérase immobilisée pour le dosage de l'acétylcholine *C.R. Acad. Sci. Paris, Ser. C*, **286**, 115-118.

[102] Tran-Minh C. and Beaux J. (1978) Etude du comportement des électrodes sensibles à l'urée en milieu fluorure. *C.R. Acad. Sci. Paris, Ser. C*, **287**, 191-194.

[103] Berjonneau A.M., Tran-Minh C. and Broun G. (1972) Mesure continue de l'activité des décarboxylases de la phénylalanine et de la tyrosine. *C.R. Acad. Sci. Paris, Ser. D*, **275**, 121-124.

[104] Guilbault G.G. and Hrabánková E. (1970) Determination of urea in blood and urine with a urea-sensitive electrode. *Anal. Chim. Acta*, **52**, 287-294.

[105] Pandey P.C., Pandey V., Tran-Minh C. and Lantreibecq F. (1992) A new sensor for the analysis of urea in human serum. *Ind. J. Technol.*, **30**, 404-408.

[106] Palleschi G., Mascini M., Martinez-Fabregas E. and Alegret S. (1988) Urea determination in human sera with an ammonium ion selective electrode made with solid inner electric contact and immobilised urease. *Anal. Letters*, **21**, 1115-1129.

[107] Nilsson H., Åkerlund A.C. and Mosbach K. (1973) Determination of glucose urea and penicillin using enzyme-pH electrodes. *Biochim. Biophys. Acta*, **320**, 529-534.

[108] Tor R. and Freeman A. (1986) New enzyme membrane for enzyme electrodes. *Anal. Chem.*, **58**, 1042-1046.

[109] Kulys J.J., Gureviciene V.V., Laurinavicius V.A. and Jonuska A.V. (1986) Urease sensors based on differential antimony electrodes. *Biosensors*, **2**, 35-44.

[110] Anfält T., Graneli A. and Jagner D. (1973) A urea electrode based on the ammonia probe. *Anal. Letters*, **6**, 969-975.

[111] Papastathopoulos D.S. and Rechnitz G.A. (1975) A urea sensing membrane electrode for whole blood. *Anal. Chim. Acta*, **79**, 18-25.

[112] Mascini M. and Guilbault G.G. (1977) Urease coupled ammonia electrode for urea determination in blood serum. *Anal. Chem.*, **49**, 795-798.

[113] Mascini M. and Palleschi G. (1983) A flow through detector for simultaneous determinations of glucose and urea in serum samples. *Anal. Chim. Acta*, **145**, 213-217.

[114] Tran-Minh C. (1971) *Transfert et génération des gaz par les membranes. Applications aux générateurs et aux capteurs.* PhD Thesis, Rouen, p. 212-213.

[115] Wingard L.B., Schiller J.G., Wolfson S.R. Jr., Liu C.C., Drash A.L. and Yao S.J. (1979) Immobilized enzyme electrodes for the potentiometric measurement of glucose concentration: immobilization techniques and materials. *J. Biomed. Mat. Res.*, **13**, 921-935.

[116] Guilbault G.G. and Hrabánková E. (1970) An electrode for determination of amino acids. *Anal. Chem.*, **42**, 1779-1783.

[117] Guilbault G.G. and Hrabánková E. (1971) New enzyme electrode probes for D-amino acids and asparagine. *Anal. Chim. Acta*, **56**, 285-290.

[118] Guibault G.G. and Shu F.R. (1971) An electrode for determination of glutamine. *Anal. Chim. Acta*, **56**, 333-338.

[119] Arnold M.A. and Rechnitz G.A. (1980) Comparison of bacterial, mitochondrial, tissue and enzyme biocatalysts for glutamine selective membrane electrodes. *Anal. Chem.*, **52**, 1170-1174.

[120] Festaz-Furet B., Healy, J.C. and Tran-Minh C. (1990) Réacteur biocatalysé et procédé de traitement correspondant, applicables notamment à la transformation du vin et à la fermentation malo-lactique. F. *patent* 9 013 131; PCT Fr 91/00802.

[121] White W.C. and Guilbault G.G. (1978) Lysine specific enzyme electrode for determination of lysine in grains and foodstuffs. *Anal. Chem.*, **50**, 1481-1486.

[122] Weissbach F., Kreibich G., Bartels K. and Schülke W. (1988) Rapid determination of L-lysine with an enzyme electrode by steady-state and kinetic measurement. *Acta Biotechnol.*, **8**, 269-274.

[123] Hsiung C.P., Kuan S.S. and Guilbault G.G. (1977) A specific enzyme electrode for L-phenylalanine. *Anal. Chim. Acta*, **90**, 45-49.

[124] Fung K.W., Kuan S.S., Sung H.Y. and Guilbault G.G. (1979) Methionine selective enzyme electrode. *Anal. Chem.*, **51**, 2319-2324.

[125] Walters R.R., Johnson P.A. and Buck R.P. (1980) Histidine ammonia-lyase enzyme electrode for determination of L-histidine. *Anal. Chem.*, **52**, 1684-1690.

[126] Fatibello-Filho O., Suleiman A.A., Guilbault G.G. and Lubrano G.J. (1988) Bienzymatic electrode for the determination of aspartame in dietary products. *Anal. Chem.*, **60**, 2397-2399.

[127] Llenado R.A. and Rechnitz G.A. (1971) Improved enzyme electrode for amygdalin. *Anal. Chem.*, **43**, 1457-1461.

[128] Meier H., Kumaran S., Danna A.M. and Tran-Minh C. (1991) Rapid measurement of penicillin contained in complex media using enzyme loaded glass electrodes. *Anal. Chim. Acta*, **249**, 405-411.

[129] Deng I. and Enke C. (1980) Adenosine-selective electrode. *Anal. Chem.*, **52**, 1937-1940.

[130] Bradley C.R. and Rechnitz G.A. (1984) Kinetic analysis of enzyme electrode response. *Anal. Chem.*, **56**, 664-667.

[131] Tran-Minh C. and Guyonnet R. (1978) Capteur enzymatique pour la détection et le dosage en continu des substances toxiques. *C.R. Acad. Sci. Paris, Ser. C*, **286**, 357-359.

[132] Tran-Minh C. and Beaux J. (1979) Enzyme electrodes for inhibitors determination: urease-fluoride system. *Anal. Chem.*, **51**, 91-95.

[133] Kumaran S. and Tran-Minh (1992) Insecticide determination with enzyme electrodes using different enzyme immobilization techniques. *Electroanalysis*, **4**, 949-954.

[134] Guyonnet R. (1978) *Etude des interactions entre enzymes immobilisées et inhibiteurs au moyen des radiotraceurs. Application à la réalisation d'électrodes enzymatiques pour le dosage des composés toxiques.* PhD thesis, Lyon.

[135] Tran-Minh C. and El Yamani H. (1988) Réalisation d'un capteur de toxicité. *Report*, n° 87967, Agence de l'Eau RMC, Armines.

[136] Hô M.H. (1986) Immobilized enzyme probes based on fluoride ion detection. *Proc. 2nd Int. Meeting on chemical sensors*, Bordeaux, 630-633.

[137] Booker H.E. and Haslam J.L. (1974) Immobilized enzyme probe for the determination of arginase. *Anal. Chem.*, **46**, 1054-1059.

[138] Rusling J.F., Luttrell G.H., Cullen L.F. and Papariello G.J. (1976) Immobilized enzyme-based flowing-stream analyser for measurement of penicillin in fermentation broths. *Anal. Chem.*, **48**, 1211-1215.

[139] Racine P. and Mindt W. (1971) On the role of substrate diffusion in enzyme electrodes. Biological aspects of electrochemistry. *Experimentia Suppl.*, **18**, 525-534.

[140] Mell L.D. and Maloy J.T. (1976) Amperometric response enhancement of the immobilized glucose oxidase enzyme electrode. *Anal. Chem.*, **48**, 1597-1601.

[141] Bergel A. and Comtat M. (1984) Theoretical evaluation of transient responses of an amperometric enzyme electrode. *Anal. Chem.*, **56**, 2904-2908.

[142] Notin M., Guillien R. and Nabet P. (1972) Le dosage du glucose sanguin à l'aide d'une électrode à enzyme. *Ann. Biol. Clin.*, **30**, 193-197.

[143] Romette J.L., Froment F. and Thomas D. (1979) Glucose oxidase electrode. Measurement of glucose in samples exhibiting high variability in oxygen content. *Clin. Chim. Acta*, **95**, 249-253.

[144] Su Y.C. and Chen C.Y. (1987) A glucose sensor for fermentation process monitoring: sensor fabrication and its batch operation characteristics. *Proc. Natl. Sci. Counc.* B. ROC., **11**, 10-17.

[145] Guilbault G.G. and Lubrano G.J. (1972) Enzyme electrode for glucose. *Anal. Chim. Acta*, **60**, 254-256.

[146] Thévenot D.R., Sternberg R., Coulet P.R., Laurent J. and Gautheron D. (1979) Enzyme collagen membrane for electrochemical determination of glucose. *Anal. Chem.*, **51**, 96-100.

[147] Yao T. (1983) A chemically modified enzyme membrane electrode as an amperometric glucose sensor. *Anal. Chim. Acta*, **148**, 27-33.

[148] Watanabe T., Okawa Y., Tsuzuki H., Yoshida S. and Nikei Y. (1988) Glucose oxidase-modified SnO_2 electrode as electrochemical glucose sensor. *Chem. Letters*, 1183-1186.

[149] Sternberg R., Bindra D.S., Wilson G.S. and Thevenot D.R. (1988) Covalent enzyme coupling on cellulose acetate membranes for glucose sensor development. *Anal. Chem.*, **60**, 2781-2786.

[150] Feistel C.C., Sternburg C.F., Luther C.D. and Gayleard L.L. (1988) Design considerations of an immobilized enzyme electrode for measurement of glucose in whole blood. *NATO ASI Ser., Ser. C*, **226**, 341-358.

[151] Genshaw M.A. and Jones J.E. (1989) Whole blood glucose enzyme electrode. *J. Electrochem. Soc.*, **136**, 414-417.

[152] Fischer U. and Abel P. (1982) A membrane combination for implantable glucose sensors. Measurements in undiluted biological fluids. *Trans. Am. Soc. Artif. Int. Organs*, **28**, 245-248.

[153] Rispon J., Shabtai Y., Rosen I., Zibenberg Y., Tor R. and Freeman A. (1990) In situ glucose monitoring in fermentation broth by "Sandwiched" glucose oxidase electrode (SGE). *Biotech. Bioeng.*, **35**, 103-107.

[154] Kasuno H. (1989) Glucose enzyme electrode with percutaneous interface which operates independently of dissolved oxygen. *Clin. Phys. Physiol. Meas.*, **10**, 1-9.

[155] Enfors S.O. (1987) Compensated enzyme-electrodes for in-situ process control, in *Biosensors. Fundamentals and applications.* (ed. A.P.F. Turner, I. Karube and G.S. Wilson), Oxford Univ. Press, 347-355.

[156] Durliat H. and Comtat M. (1984) Amperometric enzyme electrode for determination of glucose based on thin layer spectroelectrochemistry of glucose oxidase. *Anal. Chem.*, **56**, 148-152.

[157] Mahenc J. and Aussaresses H. (1979) Electrode à enzyme spécifique du glucose basée sur la détection ampérométrique de l'hexacyanoferrate II. *C.R. Acad. Sci. Paris, Ser. C*, **289**, 357-359.

[158] Cass A.E.G., Davis G., Francis G.D., Hill H.A.O., Aston W.J., Higgins I.J., Plotkin E.V., Scott L.D.L. and Turner A.P.F. (1984) Ferrocene-mediated enzyme electrode for amperometric determination of glucose. *Anal. Chem.*, **56**, 667-671.

[159] Brooks S.L., Ashby R.E., Turner A.P.F., Calder M.R. and Clarke D.J. (1987/88) Development of an on-line glucose sensor for fermentation monitoring. *Biosensors*, **3**, 45-46.

[160] Kulys J.J., Samalins A.S. and Svirmickas G.J.S. (1980) Electron exchange between the enzyme active center and organic metal. *FEBS Letters*, **114**, 7-10.

[161] Boutelle M.G., Stanford C., Fillenz M., Albery W.J. and Bartlett P.N. (1987) An amperometric enzyme electrode for continuous monitoring of brain glucose in the freely moving rat. *Neurosci. Letters*, **72**, 283-288.

[162] Albery W.J. and Craston D.H. (1987) Amperometric enzyme electrodes: Theory and experiment, in *Biosensors. Fundamentals and applications*. (ed. A.P.F. Turner, I. Karube and G.S. Wilson), Oxford Univ. Press, 180-210.

[163] Kulys J.J. (1986) Enzymes electrodes based on organic metals. *Biosensors*, **2**, 3-13.

[164] Mullen W.H., Churchouse S.J. and Vadgama P.M. (1985) Enzyme electrode for glucose based on the quinoprotein glucose dehydrogenase. *Analyst.*, **110**, 925-928.

[165] Coulet P.R. and Bertrand C. (1979) Asymmetrical coupling of enzymatic systems on collagen membranes. Application to multienzyme electrodes. *Anal. Letters*, **12**, 581-587.

[166] Pfeiffer D., Scheller F., Janchen M., Bertermann K. and Weise H. (1980) Bienzyme electrodes for ATP, NAD^+, starch and disaccharides based on a glucose sensor. *Anal. Letters*, **13**, 1179-1200.

[167] Clark L.C. Jr. (1972) A family of polarographic enzyme electrodes and the measurement of alcohol. *Biotech. Bioeng. Symp.*, **3**, 377-394.

[168] Suzuki S., Takahashi F., Satoh I. and Sonobe N. (1975) Ethanol and lactic acid sensors using electrodes coated with dehydrogenase-collagen membranes. *Bull. Chem. Soc. Japan*, **48**, 3246-3249.

[169] Blaedel W.J. and Engstrom R.C. (1980) Reagentless enzyme electrodes for ethanol, lactate and malate. *Anal. Chem.*, **52**, 1691-1697.

[170] Verduyn C., Van Dijken J.P. and Scheffers W.A. (1983) A simple, sensitive, and accurate alcohol electrode. *Biotech. Bioeng.*, **25**, 1049-1055.

[171] Kjellén K.G. and Neujahr H.Y. (1980) Enzyme electrode for phenol. *Biotech. Bioeng.*, **22**, 299-310.

[172] Bertrand C., Coulet P.R. and Gautheron D.C. (1979) Enzyme electrode with collagen-immobilized cholesterol oxidase for the microdetermination of free cholesterol. *Anal. Letters*, **12**, 1477-1488.

[173] Durliat H., Comtat M. and Mahenc J. (1975) Dosage ampérométrique du lactate à l'aide d'une électrode à enzyme. *J. Electroanal. Chem.*, **66**, 73-76.

[174] Mizutani F., Yamanaka T., Tanabe Y. and Tsuda K. (1985) An enzyme electrode for L-lactate with a chemically amplified electrode. *Anal. Chim. Acta*, **177**, 153-166.

[175] Mascini M., Fortunati S., Moscone D., Palleschi G., Massi-Benedetti M. and Fabietti P. (1985) An L-lactate sensor with immobilized enzyme for use in *in vivo* studies with an endocrine artificial pancreas. *Clin. Chem.*, **31**, 451-453.

[176] Bardeletti G., Séchaud F. and Coulet P.R. (1986) A reliable L-lactate electrode with a new membrane for enzyme immobilization for amperometric assay of lactate. *Anal. Chim. Acta*, **187**, 47-54.

[177] Mascini M., Mazzei F., Moscone D., Calabrese G. and Massi-Benedetti M. (1987) Lactate and pyruvate electrochemical biosensors for whole blood in extracorporeal experiments with an endocrine artificial pancreas. *Clin. Chem.*, **33**, 591-593.

[178] Biade A.E., Bourdillon C., Laval J. M., Mairesse G. and Moiroux J. (1992) Complete conversion of L-lactate into D-lactate. A generic approach involving enzymatic catalysis, electrochemical oxidation of NADH and electrochemical reduction of pyruvate. *J. Am. Chem. Soc.*, **114**, 893-897.

[179] Romette J.L., Yang J.S., Kasukabe H. and Thomas D. (1983) Enzyme electrode for specific determination of L-lysine. *Biotech. Bioeng.*, **25**, 2557-2566.

[180] Kubo I., Karube I. and Suzuki S. (1983) Amperometric determination of creatinine with a sensor based on immobilized creatininase and nitrifying bacteria. *Anal. Chim. Acta*, **151**, 371-376.

[181] Renneberg R., Pfaffer D., Scheller F. and Janchen M. (1982) Enzyme sequence and competition electrodes based on immobilized glucose oxidase, peroxidase and catalase. *Anal. Chim. Acta*, **134**, 359-364.

[182] Comtat M., Galy M., Goulas P. and Souppe J. (1988) Amperometric bienzyme electrode for L-carnitine. *Anal. Chim. Acta*, **208**, 295-300.

[183] Matsumoto K., Seijo H., Karube I. and Suzuki S. (1980) Amperometric determination of choline with use of immobilized choline oxidase. *Biotech. Bioeng.*, **22**, 1071-1086.

[184] Aizawa M., Karube I. and Suzuki S. (1974) A specific bioelectrochemical sensor for hydrogen peroxide. *Anal. Chim. Acta*, **69**, 431-437.

[185] Guilbault G.G. and Nanjo M. (1975) A phosphate selective electrode based on immobilized alkaline phosphatase and glucose oxidase. *Anal. Chim. Acta*, **78**, 69-80.

[186] Bergveld P. (1985) The impact of MOSFET based sensors. *Sensors Actuators*, **8**, 109-127.

[187] Miyahara Y., Maruzumi T., Shiokawa S., Matsuoka H., Karube I. and Suzuki S. (1983) Micro urea sensor using semiconductor and enzyme immobilizing technologies. *Chem. Soc. Japan*, **6**, 823-830.

[188] Gardies F., Jaffrezic-Renault N., Martelet C, Perrot H., Valleton J.M. and Alegret S. (1990) Microenzyme field effect transistor sensor using direct covalent bonding of urease. *Anal. Chim. Acta*, **231**, 305-308.

[189] Eddowes M.J., Pedley D.G. and Webb B.C. (1985) Response of an enzyme-modified pH-sensitive ion-selective device; experimental study of a glucose oxidase-modified ISFET in buffer and unbuffer aqueous solution. *Sensors Actuators*, **7**, 233-244.

[190] Hanazato Y., Inatomi K.I., Nakako M., Shiono S. and Maeda M. (1988) Glucose-sensitive field-effect transistor with a membrane containing co-immobilized gluconolactonase and glucose oxidase. *Anal. Chim. Acta*, **212**, 49-59.

[191] Winquist F., Spetz A. and Lundström I. (1984) Determination of urea with an ammonia gas-sensitive semi-conductor device in combination with urease. *Anal. Chim. Acta*, **163**, 143-149.

[192] Tamiya E., Seki A., Karube I., Gotoh M. and Shimizu I. (1988) Hypoxanthine sensor based on an amorphous silicon FET. *Anal. Chim. Acta*, **215**, 301-305.

[193] Tamiya E., Karube I., Kitagawa Y., Ameyama M. and Nakashima K. (1988) Alcohol FET sensor based on a complex cell membrane enzyme system. *Anal. Chim. Acta*, **207**, 77-84.

[194] Gotah M., Tamiya E. and Karube I. (1989) Micro-FET biosensors using polyvinylbutyral membrane. *J. Membrane Sci.*, **41**, 291-303.

[195] Kelly R.G. and Owen A.E. (1985) Microelectronic ion sensors: a critical survey. *IEE Proceedings*, **132**, 227-236.

[196] Van der Schoot B.H. and Bergveld P. (1987/88) ISFET based enzyme sensors. *Biosensors*, **3**, 161-186.

[197] Peterson J.I., Goldstein S.R., Fitzgerald R.Y.and Buckhold D.K. (1980) Fiber optic pH probe for physiological use. *Anal. Chem.*, **52**, 864-869.

[198] Grattan K.T.V., Mouaziz Z. and Palmer A.W. (1987) Dual wavelength optical fibre sensor for pH measurement. *Biosensors*, **3**, 17-25.

[199] Kirkeight G.F., Narayanaswamy R. and Welti N.A. (1984) Fibre optic pH probe based on the use of an immobilized calorimetric indicator. *Analyst*, **109**, 1025-1028.

[200] Arnold M.A. and Ostler T.J. (1986) Fiber optic ammonia gas sensing probe. *Anal. Chem.*, **58**, 1137-1140.

[201] Giuliani J.F., Wohltjen H. and Jarvis N.L. (1983) Reversible optical waveguide sensor for ammonia vapors. *Optics Letters*, **8**, 54-56.

[202] Seitz W.R. (1984) Chemical sensors based on fiber optics. *Anal. Chem.*, **56**, 16A-34A.

[203] Zhujun Z. and Seitz W.R. (1984) A carbon dioxide sensor based on fluorescence. *Anal. Chim. Acta*, **160**, 305-309.

[204] Wolfbeis O.S., Weis L.J., Leiner M.J.P. and Ziegler W.E. (1988) Fiber-optic fluorosensor for oxygen and carbon dioxide. *Anal. Chem.*, **60**, 2028-2030.

[205] Walt D.R., Munkholm C., Yuan P., Luo S. and Barnard S. (1989) Design, preparation, and applications of fiber-optic chemical sensors for continuous monitoring. *A.C.S. Symposium Series*, **403**, 253-272.

[206] Arnold M.A. (1989) Fiber-optic based biocatalytic biosensors. *A.C.S. Symposium Series*, **403**, 303-317.

[207] Wolfbeis O.S. (1986) Fiber-optic probe for kinetic determination of enzyme activities. *Anal. Chem.*, **58**, 2874-2876.

[208] Wangsa J. and Arnold M.A. (1988) Fiber-optic biosensors based on the fluorometric detection of reduced nicotinamide adenine dinucleotide. *Anal. Chem.*, **60**, 1080-1082.

[209] Mottin S., Tran-Minh C., Laporte P., Cespuglio R. and Jouvet M. (1992) Fiber-optic time resolved fluorescent sensor for in vivo measurements in the brain of rat unanesthetized and freely moving. *14th Annual Int. Conf. IEEE, in Proc. Sat. Symp. on Neurosciences and Technol.* (ed. A. Dittmar and J.C. Froment), Lyon, 235-238.

[210] Jordan D.M., Walt D.R. and Milanovich F.P. (1987) Physiological pH fiber-optic chemical sensor based on energy transfer. *Anal. Chem.*, **59**, 437-439.

[211] Peterson J.I., Fitzgerald R.V. and Buckhold D.K. (1984) Fiber optic probe for *in vivo* measurement of oxygen partial pressure. *Anal. Chem.*, **56**, 62-67.

[212] Lee E.D., Werner T.C. and Seitz R.W. (1987) Luminescence ratio indicators for oxygen. *Anal. Chem.*, **59**, 279-283.

[213] Lübbers D.W. and Opitz N. (1983) Optical fluorescence sensors for continuous measurement of chemical concentrations in biological systems. *Sensors Actuators*, **4**, 641-654.

[214] Wyatt W.A., Poirier G.E., Bright F.V. and Hieftje G.M. (1987) Fluorescence spectra and lifetimes of several fluorophores immobilized on nonionic resins for use in fiber-optic sensors. *Anal. Chem.*, **59**, 572-576.

[215] Schultz J.S. (1982) Optical sensor for plasma constituents. US *patent* 4344438.

[216] McCapra F. (1987) Potential applications of bioluminescence and chemiluminescence in biosensors, in *Biosensors. Fundamentals and applications.* (ed. A.P.F. Turner, I. Karube and G.S. Wilson), Oxford Univ. Press, 617-637.

[217] Jablonski E.G. and De Luca M. (1982) Analytical applications of bioluminescence: marine bacterial system, in *Clinical and biochemical luminescence.* (ed. L.J. Kricka and T.J.N. Carter), Dekker, New York, 75-87.

[218] Aizawa M., Ikariyama Y. and Kuno H. (1984) Photovoltaic determination of hydrogen peroxide with a biophotodiode. *Anal. Letters*, **17**, 555-564.

[219] Mattiasson B.O., Danielsson B. and Mosbach K. (1976) Enzyme thermistor assay of cholesterol, glucose, lactose and uric acid in standard solutions as well as in biological samples. *Anal. Letters*, **9**, 217-234.

[220] Fulton S.P., Cooney C.L. and Weaver J.C. (1980) Thermal enzyme probe with differential temperature measurements in a laminar flow-through cell. *Anal. Chem.*, **52**, 505-508.

[221] Rich S., Ianniello R.M. and Jespersen N.D. (1979) Development and application of a thermistor enzyme probe in the urea-urease system. *Anal. Chem.*, **51**, 204-206.

[222] Muchlbauer M.J., Guilbeau E.J. and Towe B.C. (1989) Model for a thermoelectric enzyme glucose sensor. *Anal. Chem.*, **61**, 77-83.

[223] Guilbault G.G. and Luong J.H. (1988) Gas phase biosensors. *J. Biotechnol.*, **9**, 1-10.

[224] Hô M.H., Guilbault G.G. and Scheide E.P. (1982) Detection of carbon monoxide in ambient air with a piezoelectric crystal. *Anal. Chem.*, **54**, 1998-2002.

[225] Guilbault G.G., Affolter J., Tomita Y. and Kolesar Jr. E.S. (1981) Piezoelectric crystal coating for detection of organophosphorous compounds. *Anal. Chem.*, **53**, 2057-2060.

[226] Lai C.S.I., Moody G.J., Thomas J.D.R., Mulligan D.C., Stoddart J.F. and Zarzycki R. (1988) Piezoelectric quartz crystal detection of benzene vapour using chemically modified cyclodextrins. *J. Chem. Soc. Perkin Trans.*, **2**, 319-324.

[227] Muramatsu H., Tamiya E., Suzuki M. and Karube I. (1989) Quartz crystal gelation detector for the determination of fibrinogen concentration. *Anal. Chim. Acta*, **217**, 321-326.

[228] Guilbault G.G., Lubrano G., Ngeh J., Jordan J.M. and Foley P. (1986) A protein coated piezoelectric crystal detector, SMCCR-RSB, p. 1-40.

[229] Guilbault G.G. and Ngeh-Ngwainbi J. (1988) Use of protein coatings on piezoelectric crystals for assay of gaseous pollutants. *NATO ASI Ser., Ser. C*, **226**, 187-194.

[230] Guilbault G.G. and Jordan J.M. (1988) Analytical uses of piezoelectric crystals: a review. *CRC Critical Reviews in Analytical Chemistry*, **19**, 1-28.

[231] Hô M.H. (1986) Chemical and biochemical sensors using piezoelectric transducer. *Proc. 2nd Int. Meeting on chemical sensors*. Bordeaux, 639-643.

[232] Valleton J.M. (1987) Rapport sur les avancées récentes dans le domaine des biocapteurs, AEPA 83/804, p. 18.

[233] Hikuma M., Kubo T., Yasuda T., Karube I. and Suzuki S. (1979) Microbial electrode sensor for alcohols. *Biotech. Bioeng.*, **21**, 1845-1853.

[234] Neujahr H.Y. and Kjellén K.G. (1979) Bioprobe electrode for phenol. *Biotech. Bioeng.*, **21**, 671-678.

[235] Wollenberger U., Scheller F. and Atrat P. (1980) Microbial membrane electrode for the determination of cholesterol. *Anal. Letters*, **13**, 825-836.

[236] Planchard A., Mignot L. and Junter G.A. (1988) Microbial sensor based on lipoic acid reduction. *Sensors Actuators*, **14**, 9-17.

[237] Hikuma M., Obana H. and Yasuda T. (1980) A potentiometric microbial sensor based on immobilized *Escherichia coli* for glutamic acid. *Anal. Chim. Acta*, **116**, 61-67.

[238] Kobos R.K. and Rechnitz G.A. (1977) Regenerable bacterial membrane electrode for L-aspartate. *Anal. Letters*, **10**, 751-758.

[239] Jensen M.A. and Rechnitz G.A. (1978) Bacterial membrane electrode for L-cystéine. *Anal. Chim. Acta*, **101**, 125-130.

[240] Rechnitz G.A., Riechel T.L., Kobos R.K. and Meyerhoff M.E. (1978) Glutamine selective membrane electrode that uses living bacterial cells. *Science*, **199**, 440-441.

[241] Di Paolantonio C.L., Arnold M.A. and Rechnitz G.A. (1981) Serine-selective membrane probe based on immobilized anaerobic bacteria and a potentiometric ammonia gas sensor. *Anal. Chim. Acta*, **128**, 121-127.

[242] Solsky R.L. and Rechnitz G.A. (1979) Antibody-selective membrane electrodes. *Science*, **204**, 1308-1389.

[243] Bush D.L. and Rechnitz G.A. (1987) Antibody response of polymer membrane electrodes incorporating antigenic ionophores. *J. Membrane Sci.*, **30**, 313-322.

[244] D'Orazio P. and Rechnitz G.A. (1977) Ion-electrode measurements of complement and antibody levels using marker-loaded sheep red blood cell ghosts. *Anal. Chem.*, **49**, 2083-2086.

[245] Umezawa Y., Shiba K., Watanabe T., Ogawa S. and Fujiwara S. (1980) A microimmunoelectrode. *3rd Symposium on ion-selective electrodes*, Matrafüred, 344-357.

[246] Tromberg B.J., Sepaniak M.J., Vo Dinh T. and Griffin G.D. (1987) Fiber-optic chemical sensors for competitive finding fluoroimmunoassay. *Anal Chem.*, **59**, 1226-1230.

[247] Kooyman R.P.H., Kolkman H. and Greve J. (1987) Immunoreactions monitored by detection of surface-plasmons, in *Biosensors international workshop*. (ed. R.D. Schmid *et al.*) GBF Monographs, vol. 10, 295-296.

[248] Collins S. and Janata J. (1982) A critical evaluation of the mechanism of potential response of antigen polymer membranes to the corresponding antiserum. *Anal. Chim. Acta*, **136**, 93-99.

[249] Gardies F., Martelet C., Colin B. and Mandrand B. (1989) Feasibility of an immunosensor based upon capacitance measurements. *Sensors Actuators*, **17**, 461-464.

[250] Billard V., Martelet C., Binder P. and Thérasse J. (1991) Toxine detection, using capacitance measurements on immunospecies grafted onto a semiconductor substrate. *Anal. Chim. Acta*, **249**, 367-372.

[251] Kanazawa K.K. and Gordon II J.G. (1985) The oscillation frequency of a quartz resonator in contact with a liquid. *Anal. Chim. Acta*, **175**, 99-105.

[252] Thompson M., Arthur C.L. and Dhaliwal G.K. (1986) Liquid-phase piezoelectric and acoustic transmission studies of interfacial immunochemistry. *Anal. Chem.*, **58**, 1206-1209.

[253] Davis K.A. and Leary T.R. (1989) Continuous liquid-phase piezoelectric biosensor for kinetic immunoassays. *Anal. Chem.*, **61**, 1227-1230.

[254] Muramatsu H., Tamiya E. and Karube I. (1989) Determination of microbes and immunoglobulins using a piezoelectric biosensor. *J. Membrane Sci.*, **41**, 281-290.

[255] Ikariyama Y., Furuki M. and Aizawa M. (1985) Sensitive bioaffinity sensor with mutastable molecular complex receptor and enzyme amplifier. *Anal. Chem.*, **57**, 496-500.

[256] Itagaki H., Hakoda Y., Suzuki Y. and Haga M. (1983) Drug sensor: an enzyme immunoelectrode for theophylline. *Chem. Pharm. Bull.* (1983), **31**, 1283-1288.

[257] Boitieux J.L., Desmet G. and Thomas D. (1979) An "antibody electrode". Preliminary report on a new approach in enzyme immunoassay. *Clin. Chem.*, **25**, 318-321.

[258] Gyss C. and Bourdillon C. (1987) Enzymatic electrocatalysis as a strategy for electrochemical detection in heterogeneous immunoassays. *Anal. Chem.*, **59**, 2350-2355.

[259] Gebauer C.R. and Rechnitz G.A. (1982) Deaminating enzyme labels for potentiometric enzyme immunoassay. *Anal. Biochem.*, **124**, 338-348.

[260] Aizawa M. (1987) Photochemical biosensors, in *Biosensors international workshop*. (ed. R.D. Schmid *et al.*) GBF Monographs, vol. 10, 217-222.

[261] Mattiasson B. (1977) A general enzyme thermistor based on specific reversible immobilization using the antigen-antibody interaction. *FEBS Letters*, **77**, 107-110.

[262] Arnold M.A. and Rechnitz G.A. (1981) Tissue based membrane electrode with high biocatalytic activity for measurement of adenosine 5'-monophosphate. *Anal. Chem.*, **53**, 1837-1842.

[263] Smit N. and Rechnitz G.A. (1984) Leaf based biocatalytic membrane electrodes. *Biotech. Letters*, **6**, 209-214.

[264] Sidwell J.S. and Rechnitz G.A. (1986) Progress and challenges for biosensors using plant tissue materials. *Biosensors*, **2**, 221-233.

[265] Belli S.L., Buch R.M. and Rechnitz G.A. (1987) Response model for chemo-receptor biosensors. *Anal. Letters*, **20**, 327-336.

[266] Ho M.Y.K. and Rechnitz G.A. (1987) Highly stable biosensor using artificial enzyme. *Anal. Chem.*, **59**, 536-537.

[267] Ruzicka J. and Hansen E.H. (1978) Flow injection analysis. Part X - Theory, techniques and trends. *Anal. Chim. Acta*, **99**, 37-76.

[268] Meier H. and Tran-Minh C. (1992) Determination of penicillin V in standard solution and in fermentation broth by flow injection analysis using fast responding enzyme glass electrodes in different detection cells. *Anal. Chim. Acta*, **264**, 13-22.

[269] Kumaran S. and Tran-Minh C. (1992) Determination of organophosphorous and carbamate insecticides by flow injection analysis. *Anal. Biochem*, **200**, 187-194.

[270] Bertermann K., Elze P., Scheller F., Pfeiffer D. and Janchen M. (1982) The application of an enzyme electrode for blood glucose determination in the automated flow system. *Anal. Letters*, **15**, 397-404.

[271] Yao T. (1983) Flow injection analysis for cholinesterase in blood serum by use of a choline-sensitive electrode as an amperometric detector. *Anal. Chim. Acta*, **153**, 169-174.

[272] Yao T., Kobayashi Y. and Sato M. (1983) Amperometric determination of phospholipides in blood serum with lecithin-selective electrode in a flow injection system. *Anal. Chim. Acta*, **153**, 337-340.

[273] Schelter-Graf A., Huck H. and Schmidt H.L. (1983) Rasche und genaue bestimmung von ethanol mittels einer oxidase-electrode in einem strömungs-injektions-system. *Z. Lebensm. Unters Forsch*, **177**, 356-358.

[274] Gorton L. and Jönsson G. (1986) A glucose electrode based on the adsorption of glucose oxidase on a palladium/gold modified carbon electrode. *J. Molecular Catalysis*, **38**, 157-159.

[275] Moody G.J., Sanghera G.S. and Thomas J.D.R. (1986) Amperometric enzyme electrode system for the flow-injection analysis of glucose. *Analyst*, **111**, 605-609.

[276] Valero F., Lafuente J., Poch M., Sola C., Aranjo A.N. and Lima J.L.F.C. (1990) On-line fermentation monitoring using flow-injection analysis. *Biotech. Bioeng.*, **36**, 647-651.

[277] Meier H., Lantreibecq F. and Tran-Minh C. (1992) Application and automation of flow injection analysis (FIA) using fast responding enzyme glass electrodes to detect penicillin in fermentation broth and urea in human serum. *J. Automat. Chem.*, **14**, 137-143.

[278] El Yamani H., Tran-Minh C., Abdul M.A. and Chavanne D. (1988) Automated system for pesticide detection. *Sensors Actuators*, **15**, 193-198.

[279] El Yamani H., Tran-Minh C., Abdul M.A. and Dupont M. (1987) Réalisation d'un ensemble automatisé pour la mesure de toxicité des eaux de rivière. *J. Français d'Hydrologie*, **18**, 67-75.

[280] Albisser A.M. (1982) Insulin delivery systems: Do they need a glucose sensor? *Diabetes Care*, **5**, 166-173.

[281] Thévenot D.R. (1982) Problems in adapting a glucose oxidase electrochemical sensor into an implantable glucose sensing device. *Diabetes Care*, **5**, 184-189.

[282] Mascini M. and Palleschi G. (1989) Design and applications of enzyme electrode probes. *Selective Electrode Reviews*, **11**, 191-264.

[283] Kress-Rogers E. and D'Costa E.J. (1985) Biosensors for the food industry. *IUPAC Congress*, Manchester.

[284] Karube I., Matsuoka H., Suzuki S., Watanabe E. and Toyama K. (1984) Determination of fish freshness with an enzyme sensor system. *J. Agric. Food Chem.*, **32**, 314-319.

[285] Watanabe E., Endo H. and Toyama K. (1988) Determination of inosine-5'-monophosphate in the presence of inosine and hypoxanthine with an enzyme sensor. *Appl. Microb. Biotechnol.*, **29**, 341-345.

[286] Adamowicz E. and Burstein C. (1987) L-lactate enzyme electrode obtained with immobilized respiratory chain from *E. coli* and oxygen probe for specific determination of L-lactate in yogurt, wine and blood. *Biosensors*, **3**, 27-43.

[287] Enfors S.O. and Nilson H. (1979) Design and response characteristics of an enzyme electrode for measurement of penicillin in fermentation broth. *Enzyme Microb. Technol.*, **1**, 260-264.

[288] Scheper T., Gebauer A., Sauerbrei A., Niehoff A. and Schügerl K. (1984) Measurement of biological parameters during fermentation process. *Anal. Chim. Acta*, **163**, 111-118.

[289] Onatzky J.P. and Ferrier G. (1986) Procédés et dispositifs pour détecter la pollution de l'eau. F. *patent* 2 573 875 (1986).

[290] Tran-Minh C. and Kumaran S. (1990) Biosensor for assessment of aquatic toxicity. *Biosensors'90*, Singapore.

[291] El Yamani H., Abdul M.A., Tran-Minh C. and Dupont M. (1988) Biosensore per misurazione di tossicita. *Technologie Chimiche*, **4**, 112-115.

[292] Tran-Minh C. (1988) Biocapteurs pour la détection des agents neurotoxiques. *Journées "Science et Défense"*. Protection contre les dangers nucléaires et chimiques, Saint-Etienne.

[293] Scheller F.W., Schubert F., Renneberg R., Müller H.G., Jänchen M. and Weise H. (1985) Biosensors: Trends and commercialization, *Biosensors*, **1**, 135-160.

[294] Guilbault G.G. (1982) Enzymatic glucose electrodes. *Diabetes Care*, **5**, 181-183.

[295] Schmid R.D. (1986) Biotechnology in Japan. *Appl. Microb. Biotechnol.*, **24**, 355-365.

Index

absorption spectrum 14
acetic bacteria 146
acetylcholine 80; 120; 164
acetylcholinesterase 34; 165; 180
  reaction 80; 87
  sensor 87; 91
activator 91
adenosine 80
ADP 31
adsorption 39
aerosols 26
albumin 92
alcohol 106
  alcoholic drinks 178
  dehydrogenase 15; 107
  FIA 173
  oxidase 106
alkaline phosphatase 111; 125
alpha-fetoprotein 157; 160; 177
amidase 77
amino acid
  biosensor 165
  electrode 75; **76**; 109; 110
  foodstuffs 178
amino acid oxidase **109**; 125
  reaction 75
  selectivity 68
ammonia 125
AMP 163
amperometry 11; 92
amygdalin 78
amylase 106
antennule 43
anti-HSA 120; 156
antibiotic 151
antibody 15; **39**; 153
antigen 15; 41; 153; 156
antimicrobial agent 21
apoenzyme 31
APTES 35
arginine 110; 149
arteriosclerosis 177
ascorbic acid 30; **69**; 99; 164
asparaginase 161
aspartame 78
aspartate 150
ATP 31; 133
automation 174; 182
avidin 158
background noise 18
bacteria 38; 178
bacteriological weapon 6
batch analysis 166
benzene 143
benzopyrene 155
bilirubin 111
bio/chemiluminescence 4
bioaffinity 41
biocatalysis 20
biocompatability 23; 176; 183
bioluminescence 14; 90; 132
biomedical sector 5
bioreceptor 7; 17; 20
  immobilization 23
biosensor
  applications 5-6
  construction 20
  historical 2-4
  implantable 5
  principles 17
biotin 158
blood 100

BOD 5; 151; 178
brain 104
BrCN 32
bromo-acetylcellulose 40
BSA 24; 37; 161
buffering capacity 71
calibration 62; 63
calibration curve 9
Callinectes sapidus 43
cancer 160
Candida albicans 158
carbamate 44; **81**; 172; 179
carbodiimide 36; 39; 42
carbon electrode 36
carbon monoxide 143
carbon paste 43; 164
carbon tetrachloride 90
cardiolipin 156
carnitine 111
catalase 30; **75**
  label 42; 160
  reaction 99; 108
cell 119
cellulose sheet 39
chemical amplification 22
chemical warfare agent 6; 87
  biosensor 180
chemiluminescence 14; 135
chloroplast 21
cholesterol 108; 150
cholinesterase 81; 125
  activity 127; 173
  electrode 174; 179
  piezoelectric sensor 144
cobratoxin 164
coenzyme 31
cofactor 3; 15; 20
  immobilization 31
  optical sensor 128
  regeneration 101; 146
collagen 38
competitive bonding 131
concanavalin A 29; 131
concentration 47
  profile 48
  range 92; 98
conducting polymer 34; 37
coreticulation 27
cortex 163
cortisol 161
coupling reaction 15
creatinin 5; 30; 111
creatinin amidohydrolase 30
cross-linking agent 25; 82
crown ether 41; 154
culture medium 39
cyclic AMP 161
cyclic potential 37; 100
cyclic voltammetry 102
cysteine 150; 163
cytochrome 108
cytochrome P450 21
daphnia 6; 90; 179
DB18C6 41
dead core 68
deaminase 77; 80
decarboxylase 68; 77; 110
defense 179
dehydrogenase 31; 128
denaturation. See enzyme
depolarization 44
detection
  chemical 14-15
  electrochemical 10
  photometric 13
  physical 9-14
  piezoelectric 13
  thermometric 12
detection cell 169
detection limit 8; 62; 83; **91**
diabetes 176
diaphorase 111
dielectric constant 164
diffusion
  coefficient 47
  rate 47
digoxin 154; 157
DIMP 143
dinitrophenol 41; 153; 161
dispersion coefficient 170

DNP 41
dopamine 164
dynamic range 100
EIA 22; 42; 158
electrode
  amperometric 92-111
  bienzyme 78
  carbon 39; 102
  glass 87
  gold 143
  multienzyme 111
  pCO2 64
  platinum 100
  pNH3 64; 74
  pNH4 70
  pO2 98
  potentiometric 59; 69-92; 120
  probe 92
  reference 44; 92
  rotating 100
  working 11
electrodeposition 38
electrolysis 101
electron transfer 102; 103
electropolymerization 34; 103
ELISA 42; 158
ENFET 116; **118**; 120
enthalpy 23; **137**; 140
enzyme 20
  active site 53; 89; 102
  activity 66; 68; 87; 127
  amplification 158
  concentration 81
  denaturation 52; 65; 85
  immobilization 2; **24**
  inhibition 53-59; 80
  kinetics 46-57
  labelling 42; 158
  loading factor 50; 52
  reactivation 86; 89; 171
  sensors 45-145
eosin 130
epoxy 41
erythrocyte 154
Escherichia coli 102; 150
ethanol 107; 178
  determination 134
  ISFET 119
  optical sensor 128
  oxidation 146
evanescent wave 13; 156
extrinsic sensor 4
FAD 3; 31; 33; 101
FAPAPP 34
FAPP 34
fermentation 3; 170; 178
fermenter 101; 182
ferricyanide 102
ferrocene 33; 37; 102
FIA. See flow injection analysis
Fick's Law 98; 104
field-effect transistor 3
fish freshness 177
fish 6; 119
flow injection analysis 164; 167-174
fluorescein isothiocyanate 127; 131; 155
fluoresceinamine 127
fluorescence 4; 14; 126; 155
  quenching 14; 129
  time-resolved 14; 129
fluoride 74
fluorocarbon 26
FMN 107; 134
food produce industry 5; 177
free boundary 68
GABA 165
galactosidase 30; 78; 106
gas electrode 11
GASFET 117
gelatine 100
glucoamylase 106
gluconic acid 74; 97
gluconolactonase 119
glucose
  analyser 181
  biomedical 176
  electrode 5; 74; 97-105

ENFET 119
FIA 173
optical sensor 130; 136
piezoelectric sensor 157
sensor market 177
thermal sensor 139
glucose oxidase
electrode 97
immobilization 24; 29; 36; 37
ISFET 119
label 42
reaction 74; 105
glutamate 127; 134; 135; 165
glutamic acid 150; 165
glutaminase 77; 125
glutamine 163
glutaraldehyde 35; 42; 82; 157
coimmobilization 30
enzyme immobilization 92
reaction 25
graphite 41
HABA 42; 158
hapten 41
HCG 153; 160; 177
heavy metals 85
hepatitis B 160; 177
hexacyanoferrate 30
histamine 165
histidine 149
hormones 164
HSA 120; 156; 159
hydrogen peroxide 100; 135; 139
hydrolase 77; 78
hypoxanthine 119; 177
hysteresis 9
IgG 155; 158; 161
immobilization
antibodies 39
antigens 41
cofactors 31
coreticulation 24
cross-linking 24
direct binding 25
electrical 37
electromagnetic 29
enzymes 2; 24
immersion 25
immunoagent 39
mediator 32
microorganisms 38
multienzymatic 29
physical 24; 38
vaporization 26
immuno-FET 120; 156
immunoagents 4; 39; 153; **156**
impedance 3; 12; 44; 120
incubation time **84**; 87; 155
inhibition
competitive 53
degree of 54
enzyme 53; 57
irreversible 84
non-competitive 56
percentage 172
reversible 81
uncompetitive 55; 83
inhibitor
determination 80-91; 171
incubation 171
inosine 177
insecticide 87; 172
instrumentation 15-17
insulin 160; 164; 176
interference 23; **66**; 102
invertase 105
ion carrier 41
ionic strength 29
ionophore 22; 154
ISFET 12; 35; 41; 117
isoelectric point 120
kinetics
first order 51; 138
zeroth order 51; 138
lactate 5; 108; 128
dehydrogenase 108; 128
monooxygenase 30
oxidase 108
lactose 30; 106

LED. See light emitting diode
light
  absorption 4; 13; 124; 126
  transmission 13
light emitting diode 16; 125
lindane 90; 179
linear region 9
liposome 154
luciferase 132
luciferin 133
luminol 135
lyase 77
lysine 78; 110
lysosome 21
magnetic bead 29
magnetic field 29
malate 77; 134
malate dehydrogenase 30
malathion 144
malolactic fermentation 77
maltose 106
mass transfer 92; 124
meat 177
mediator 3; 30; **101**; 107
  immobilization 32
membrane 27
  acetylcellulose 149
  dialysis 38
  enzymatic 48
  hydrophobic 73; 98; 146
  polyvinylbutyral 119
  prefunctionalized 28
  semi-permeable 92
mercury 85; 141; 179
metallic electrode 37
methylazinphos 91; 179
Michaelis constant 47; 97; 137
microbial population 151
microbial sensor 146-152
microorganism 4; 21; 38; **146**; 158
microsome 21
Microtox 91; 179
microtransducer 34
miniaturization 122
mitochondria 21
MOSFET 116
mutarotase 105
myocardial infarction 133
NAD 3; 31; 107
NADH 107; 134
  optical sensor 128
NADP 31
nafion 34
neostigmine 81; 84
Nernst Law 8; 10; 69
nerve fiber 43
neuroreceptor 22
neurotransmitter 44; 165; 177
nicotine 81
nitrophenol 125
NMA 33
NMP 33; 104
norepinephrine 165
nylon net 40
nystatin 151
oenology 77
olfactive mucus 165
olfactory mucosa 4
optical fiber 13; 16; 41; **123**; 155
organelle 21; 24; 43
organic conducting salt **33**; 103; 107
  mechanism 104
organic pollution 5; 178
organophosphate **87**; 143; 179
  determination 179
ornithine 110
ovalbumin 42
oxygen 98; 135; 179
  concentration 101
  sensor 130
oxygen probe 12
palladium 35
PAM 89
paraoxon 90
paraquat 91; 179
penicillin 79; 127; 178
  determination 151
  ENFET 118
  FIA 174
penicillinase 14; 79; 125

peptide 78
peroxidase 30; 111; 135
  label 161
pesticides 87; 174
pH
  effect of 64
  electrode 11; 16
  enzyme activity 83
  ISFET 117
  optical sensor 125; 126; 130
phase
  heterogeneous 57
  homogeneous 50
phenazine methosulfate 107
phenol 108; 149
phenol red 125; 130
phenylalanine 78; 110
phosphorylase nucleoside 30
photodiode 14
photolithography 36
phototransistor 125
physostigmine 84
piezoelectric
  crystal 157
  sensor 142-144
plant tissue 163
platinum black 37
pO2 electrode 14; 42
polyacrylamide 24; 100
polypyrrole 3; **34**; 37; 103
polysaccharides 105
polyvinylbutyral 41
potential 120
potentiometry 10
pregnancy 160; 177
protein 178
PVC 41
pyruvate 108; 128
quartz 13; 17; **142**; 157; 180
quinidine 154
quinolinium 33
quinone 102
rate
  constant 46
  maximal 47
reaction
  product 46
  rate 47
reactivation. See enzyme
receptor
  chemoreceptor 22; 43; 164
  immunoreceptor 22
  neuroreceptor 44
receptrode 43
reduced variables 48
REFET 120
reflectance 156
regeneration. See cofactor
renal function 177
repeatability 8
reproducibility 8
resonance frequency 158
response time 9; 49; **61**; 92
river water 175
sample throughput 169
sandwich immunoassay 161
sarcosine oxidase 30
sarin 179
selectivity 8
semiconductor 3; 12; 116
sensitivity 8; 62; 92
sensor
  amperometric 36
  calibration 62
  characteristics 7
  definition 7
  enthalpimetric 136-142
  enzyme 45-145; 176
  extrinsic 124
  immunological 153-162; 177
  integrated 122
  microbial 146-152; 177
  miniature 34
  optical 122-136; 155
  piezoelectric 142-144; 157
  potentiometric 34
  response 47-52; 61
  semiconductor 116-122; 156
  stability 52

Sepharose 32
serine 89
serum 100
Si3N4 35
signal to noise 3; 122
silanization 100
silica bead 41
silicone 27; 100
siloxane 34
SiO2 35
soman 179
spacer 32
specificity 8; 69
staphylococcus enterotoxin B 157
starch 30; 106
steady state 9; 49
sterilization 183
stimuli 44
substrate 46; 62; 168
  concentration 83
  consumption 171
sucrose 105
sulfoxide 164
surface acoustic wave 144
surface plasmon resonance 39; 156
tabun 179
TCNQ 3; 33; 103
teflon 27
TELISA 162
  temperature 65; 137
testosterone 134
theophylline 160; 177
thermal conductivity 141
thermal enzyme sensor 136-142
thermistor 12; 16; 23; **136**; 141
  reference 138
thermocouple 12; 23
thermopile 12; 136; 142
tissue 4; 21; 43; 163
toxicity 5; 179
  total 90; 179
toxin 164; 177
transducer 7; 18; 22; 66
  capacitive 164
  pCO2 71
  pH 71
  pNH3 73
  pNH4 70
transient state 47
triacetylcellulose 36
trichlorophenol 179
TTF 33; 103
tyrosine 77; 164
urea 5; 140
  electrode 70-74
  ENFET 118
urease **70**; 119; 125; 161
  immobilization 24; 35
  inhibition 85
  label 42
uric acid 30
VDRL 157
wine 5; 178
xanthine oxidase 30; 119
yeast 151
yoghurt 178

GPSR Compliance

The European Union's (EU) General Product Safety Regulation (GPSR) is a set of rules that requires consumer products to be safe and our obligations to ensure this.

If you have any concerns about our products, you can contact us on ProductSafety@springernature.com

In case Publisher is established outside the EU, the EU authorized representative is:

Springer Nature Customer Service Center GmbH
Europaplatz 3
69115 Heidelberg, Germany

Batch number: 10761599

Printed by Printforce, the Netherlands